Inspection and
Training
for TPM

Terry Wireman

Inspection and
Training
for TPM

Industrial Press Inc.

Library of Congress Cataloging-in-Publication Data

Wireman, Terry.
    Inspection and training for TPM / Terry Wireman. — 1st ed.
    256 p. 15.6 × 23.5 cm.
    Includes index.
    ISBN 0-8311-3042-3
    1. Plant maintenance — Management. 2. Industrial equipment —
Maintenance and repair — Management. 3. Industrial equipment —
Inspection. I. Title.
    TS192.W56 1992
    658.2'2—dc20   92-11059
                                                  CIP

Industrial Press Inc., 200 Madison Avenue, New York, NY 10016-4078

First Edition, First Printing

10 9 8 7 6 5 4 3 2

Contents

v

Preface

In today's world markets, most companies are looking for any competitive advantage they can find. The equation for world class competitiveness involves on-time delivery of the best quality product at the lowest possible cost.

Companies have tried many programs to solve this equation, except the most important one — maintenance. If all parts of the world class equation could have a common denominator, it is maintenance. Without properly maintained plants and facilities, it is impossible to be a world class competitor.

First, you cannot produce a quality product on equipment that is in such poor condition it cannot stay within specification. Also, you cannot deliver goods or services if the equipment or facility is unreliable and subject to breakdowns or failures. Finally, you cannot provide the lowest cost product if your maintenance costs and the resulting operational costs are higher than your competitors'. (Some studies show that over 30% of all maintenance expenditures are unnecessary.)

Maintenance is now under the corporate microscope, being examined for any possible improvements. Different programs are being implemented to try to improve maintenance and gain the related cost savings. One of the newest programs is a concept called TPM, or Total Productive Maintenance. This concept (partially used in the U.S. in the 1960's as operator-based maintenance) has been optimized by the Japanese to reduce maintenance and operational costs.

The concept is explained in the first chapter in further detail, but one of the main goals is to allow the operators to maintain and repair their own equipment. The problem is that most operators lack the technical skills necessary to perform basic maintenance tasks.

The goal of this book is to provide the basic technical information necessary for operators to maintain and repair their equipment. This book also provides the maintenance apprentices with information to increase their effectiveness. It also provides the maintenance technicians with a reference work to refresh their knowledge and skills. Finally, it will assist managers who are trying to start a TPM program, by focusing on the first step: development of a good preventive maintenance program.

Part I of the book provides an overview of a TPM program and its goals and objectives. The material then progresses into the different types of technology that may be used in the program: preventive, predictive, computerized, and nondestructive testing. Part II provides the technical information necessary to develop a good operator-based inspection and maintenance program.

It is my sincere wish that this information will help make your company more competitive in its respective marketplace.

Development and Administration of a TPM Program

1 Total Productive Maintenance

TPM — what does it mean to you? This is a question asked of countless seminar attendees. The answers are surprising, since they range from a simple preventive maintenance program to a true understanding of the Japanese concept. Since understanding TPM is important for the consideration of this text, it would be valuable to define it. TPM means Total Productive Maintenance. TPM is an equipment management program that involves all employees in the company in the maintenance and repair of the company assets, whether a facility or plant.

This definition is a departure from the past practice of each part of a company having a definite role to play with certain borders. For example, the common situation is where the operations or production department runs the equipment and the maintenance department maintains the equipment. If the equipment begins to develop a problem, the operator would call the maintenance department. The maintenance department would respond, based on its current workload and the severity of the problem the operator had identified.

If the maintenance department was working on other critical problems, and the problem the operator called in was small, it may have been a considerable time before anyone was sent to correct the problem. During this time, the problem would have grown, and either a breakdown would have occurred, or the equipment condition would have deteriorated to the point that damage had been done to the equipment or the product. In either case, additional costs were incurred.

If the equipment experienced a breakdown, then there would have been the cost of lost production, or the lost use of a portion of the facility. In addition, the cost of the maintenance action would have been increased, since equipment breakdowns are always more expensive than equipment servicing. Even if a breakdown did not occur, chances are the equipment condition would have deteriorated to a point where the quality of the product (or service) was affected. This would result in an increased quality cost. These increased costs make the company less competitive with its competitors (whether foreign or domestic).

A Total Productive Maintenance program has five definite goals, which help to change the previous situation. They are as follows.

1. Ensure equipment capacity (or facility availability).
2. Utilize a program of maintenance for the entire life of the equipment.
3. Require support from all departments involved with the use of the equipment or facility.
4. Solicit and utilize input from employees at all levels within the organization.
5. Utilize consolidated teams for continuous improvement.

Ensuring equipment capacity or the availability of the facility is a goal for all employees. If the equipment does not function properly, then the products will not be produced at the lowest cost, at the highest quality, or delivered on time. While some argue that this does not apply to facilities, what happens to employee productivity when:

the building temperature is too high? (or too low?)

enough lights are burnt out to the point where it is difficult to read?

only certain electrical receptacles actually work?

Since every employee is affected by the equipment or facility, maintenance should be the goal of everyone.

The utilization of a program for the entire life of the equipment is analogous to the preventive maintenance program. The program must go beyond just simple services to be truly effective. The maintaining activities would change as the equipment requirements changed. For example, as a car gets older, the simple lubrication checks will not be sufficient. There are major replacements, such as the brakes or the tires, the engine must be tuned, or the wheels must be aligned. Most equipment or facilities follow the same pattern. Simple or routine services will be sufficient initially (perhaps for years), but eventually the equipment wears and replacements are required. Replacing the worn components before they fail in an unplanned mode and interrupt the plant or facility operation is the real goal for all employees. This goal assures on-time delivery of the company's product or service.

Requiring support from all departments involved in the use of the plant or facility is really asking for communication. It is difficult to maintain equipment if the various groups do not communicate. For example, do you have equipment from many different manufacturers at your plant or facility that perform the same function? Consider the following questions if you have this condition.

Does it require duplication of spare parts, since they will not be interchangeable between the different manufacturers?

If the equipment was from the same manufacturer, what impact would this have on the number of inventory items currently carried in stock?

If the equipment was from the same manufacturer, wouldn't it be easier to form a strong supplier relationship, ensuring good quality and timely service from the supplier?

If the equipment is from the same manufacturer, doesn't it simplify the operational and maintenance training for the personnel involved?

If the equipment was from the same manufacturer, wouldn't it make the maintenance repair and troubleshooting easier for the technicians involved?

These questions highlight just one small area where all employees must be involved and consulted when it comes to maintaining the equipment. However, most companies still have the attitude: "we specify it (engineering); you buy it (purchasing); you run it (operations); and you maintain it (maintenance)." In this mode, the departments never really communicate until major problems develop. By then, it is too late. The employee involvement required in this step will break down many of the traditional barriers.

Utilizing the input from all employees within the organization helps to ensure that the company is maximizing the use of its resources. Many employees will think of a better way of doing things, but will keep it to themselves if they feel it is useless to bring up the idea. This calls for a change in the management styles of many companies. The dictatorial "boss–slave" managers of the past must give way to the employee empowerment programs of the present. When companies do not empower their employees, they provide their competitors a large advantage in cost, quality, and timely delivery. At present, with the competitive climate so intense, a company that has not started an empowerment program may find itself out of business quickly. Allowing the employees to find solutions to problems, to find better methods and policies, and then to implement these programs will be the charge for the management teams.

The use of consolidated teams is the combining of employees of different backgrounds and job functions into teams to work on specific problems or objectives. This will include management, operations, engineering, maintenance, purchasing, etc. These departments will communicate better and develop a company approach to business, rather than the "multiple company" or "ivory tower" approach that is so common in organizations today. The teams can be used when a problem arises or when a new policy or procedure is required to provide a product or service. This ensures that all departments "sign on" to the changes, since they had input into the change. For example, a production problem may be the target. The *operations* group would need representation to ensure the problem was properly understood and the solution was acceptable. The *maintenance* group would bring the repair and maintenance expertise, ensuring the equipment's longevity would not be endangered by the solution. The *engineering* group would be involved to ensure that the product design and quality was not compromised. The *purchasing* group may be involved if the solution required the purchase of new components or equipment to ensure the specifications and requirements were properly understood. Each group must be respected for the expertise it brings to the table. This will ensure that a competitive, cost effective solution will be provided for each decision.

The Role of Operations

What is the role of the operations department under a Total Productive Maintenance program? Under a TPM program, the operations group becomes more responsible for the care and maintenance of the equipment. The four basic objectives for the operators are as follows.

1. Perform routine services, such as cleaning, lubricating, etc.
2. Keep equipment in good operating condition by using good operational practices, visual inspections, etc.
3. Detect any deterioration or abnormalities, including wear, quality problems, etc.
4. Improve operational skills, such as optimal setups, adjustments, etc.

The operators are in the best position to perform the basic services on the equipment, such as the cleaning, basic lubrication, and adjustment of the equipment. This can be performed in a short time period each day, usually during slack time that occurs in any operation. However, in some plants, it is necessary to provide a small window (15 minutes or less) of time to perform these routine tasks.

Keeping the equipment in good operating condition is an ongoing daily task. The operator would be the first to note any abnormality in the equipment condition, anything that might be the start of a problem. This type of abnormality may be a loose fastener, a poor alignment of a component, or anything that will eventually affect the operation of the equipment if left unchecked. The solution may be tightening a bolt or replacing a cotter pin. These tasks, again, require very little of the operator's time.

Detecting abnormalities or deterioration of the equipment is closely related to the previous objective, but is more involved. The solutions to these problems may be beyond the scope of what the operator can do in a routine service period (the 15 minutes or less). It may require an equipment shutdown for component replacement. This type of service will require the involvement of the maintenance personnel to support the operator's efforts. Typical of these types of problems are: the wear is affecting quality, the equipment won't come up to full speed, or there is no further adjustment left to compensate for wear. If the operator finds these items, they can usually be corrected before any longer term problems are created.

The enhancing of the operator's skills is an ongoing training program. This ensures that the operator is always current with the latest methods and practices. This will help to provide optimal operational efficiency for the equipment. In order for this step to be effective, there will have to be strong management commitment to ongoing training for the operators of all the equipment.

WARNING

The operators will only be effective in performing these tasks in a TPM program if management commits to the training necessary to ensure that the operators know what they are doing. This is the main thrust of this book: to provide basic material to allow the operators to make good quality decisions about the condition of their equipment. *If management is not willing to invest in this type of training, DO NOT begin a TPM program;* it will only end in failure. In addition, if management begins task transfer without proper training, lawsuits will result if any operators are injured or killed while performing maintenance tasks.

The Role of Maintenance

The role of the maintenance department under a TPM program evolves to a more technical function than in a conventional organization. The objectives of the maintenance department will include:

1. providing support and training for the operators
2. performing overhauls on equipment when component replacement or more complex maintenance is required
3. identifying solutions to chronic problems by advanced diagnostics

4. continuously improving maintenance practices and methods.

Providing support and training for the operators includes assisting the operators in solving problems that are technically beyond the operator's knowledge or ability. It also includes providing the operators with the necessary skills training to complete any desired maintenance activity. The maintenance technician may provide basic skills training, equipment- or component-specific training, or even some conceptual or theoretical training if it is required to perform some operator maintenance task.

The maintenance department will also provide additional support during maintenance activities on the equipment, and in some cases will perform all of the work on major overhauls or outages. Many companies will have the operations personnel work during these outages as assistant technicians with the maintenance personnel. This will also help the training effort, thus ensuring that the operator learns more about the equipment.

The use of advanced techniques or diagnostics eventually becomes the primary focus for the maintenance department. It will use predictive techniques, supplemented by analysis of data from a computerized maintenance management system, to detect and correct chronic problems. By analysis of data, it can provide cost effective solutions, which not only makes the company more profitable, but also solves nagging problems, thus increasing the morale among the operations personnel.

The continuous improvement in maintenance practices and techniques means the maintenance personnel will stay trained in the latest management philosophies, continually looking for anything that may give a competitive edge to their practices. Beyond the philosophy is the technology. The maintenance department must stay attuned to the latest advances, looking for cost effective solutions and improvements for the plant or facility. This does not mean purchasing technology for technology's sake, but using the tools necessary to increase profitability.

In summary, under a TPM program, much of the routine maintenance work is given to the operators, while the maintenance department focuses on the more technical aspects of the equipment maintenance.

Organizational Structures

One of the most common questions is: "how should the organization be structured?" There are several philosophies that are currently being used, so each must be analyzed for the application to a company. In many cases, a combination of various aspects of the different philosophies may be used.

One school of thought is to have the maintenance personnel assigned directly to the line or operations personnel. This can work; however, the maintenance organization will still have to exist with a core of skilled personnel with a technical focus. The organizational structure may be viewed as in Figure 1-1. This structure allows the maintenance department to have some personnel for more technical assignments, but keeps the primary maintenance personnel on the team where they can be focused. This structure is really just a slight modification of the common area maintenance organization pictured in Figure 1-2. The major point is to keep the maintenance personnel focused on the maintenance responsibilities. If they are viewed as just another team member and used as another operator, the ultimate goal of improving the equipment utilization will not be achieved. A second danger in this arrangement is that some operations managers have used it as a method of getting control of the maintenance personnel. If this happens, some short term benefits appear to be gained; but in the long term, the organization suffers, since the focus of maintaining the equipment is lost.

A second school of thought is to use an area maintenance concept with maintenance supporting the operations team, but with a direct report to the central maintenance organization. This is still, in this author's experience, the most effective manner to establish the organization. The structure would appear as in Figure 1-3. There are the benefits of the operations and maintenance personnel working closely, but there is no temptation to misuse the maintenance personnel. This will undoubtedly be the primary structure used to begin a TPM program.

The Role of Management

Management's role also undergoes changes in a TPM program. Instead of being viewed as a boss, a manager becomes a coach. This

TPM
Organizational Structure

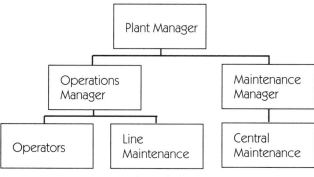

Figure 1-1.

TPM
Organizational Structure

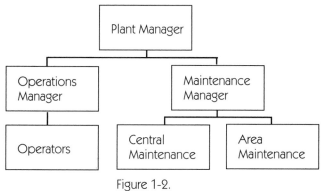

Figure 1-2.

TPM
Organizational Structure

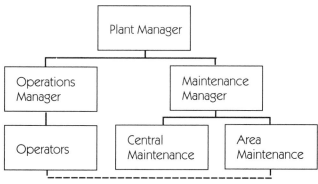

Figure 1-3.

role is difficult for many managers to develop. Some, particularly those grounded in the military, are used to giving orders on exactly how to do things. A TPM program requires a manager to "steer" the organization, but not tell everyone exactly how to do the job. The increased flexibility an employee has results in more ideas and suggestions on how to accomplish his or her job. This will eventually produce cost effective and competitive solutions to many problems.

One note, however: many companies have the mistaken impression that employee empowerment is permissive management. This is not the case. Managers are still responsible for the performance of the organization. They must still make decisions and provide direction. Otherwise, the performance of the organization will suffer instead of improve.

Initiating a TPM Program

How does a TPM program evolve? The term "evolve" is used purposely, since the development and implementation of a TPM program is a 3 to 5 year project. It requires this length of time because the project will require the changing of philosophies and the removal of many traditional barriers between operations and maintenance. In addition, it will require changes in other departments, such as engineering and purchasing.

The starting point for beginning the program is an assessment of the current status of the organization. The analogy is similar to the building of a bridge. It is difficult to build the bridge correctly if you don't know the location of the starting bank (your current status) and the location of the other side of the river (your optimum status). If these parameters are not established, your bridge will be too long (the project will be too expensive and take too long) or your bridge will be too short (the project will not meet its objectives). In either case, the project will be a failure.

Organizational Assessment

The organizational assessment used to be called a maintenance audit, but that is really a misnomer. Since maintenance is a service organization, many of the attitudes, ideas, and methodologies are influenced by other parts of the organization. Any influencing factors must be taken into consideration while assessing the condition of the maintenance organization. It is truly an examination of the attitudes of the company toward maintenance. Some of the following indicators should be used to help benchmark the maintenance organization.

Indicator	Goal
Equipment delays that are maintenance related	Less than 5%
Supervisor-to-craftworker ratio	1 to 10–15
Planner-to-craftworker ratio	1 to 10–20
Labor productivity	Greater than 60%
Planning efficiency	
labor	Greater than 90%
materials	Greater than 90%
Scheduling efficiency (weekly)	
labor	Greater than 90%
materials	Greater than 90%
Percentage of overtime	Less than 5%
Size of craft backlogs	2 to 4 weeks
Service level of the stores	Greater than 95%
Equipment availability	Greater than 95%
Percentage of maintenance work covered by work orders	Must be 100%
Percentage of planned man hours compared to total man hours	Greater than 90%
Preventive maintenance cost compared to total maintenance costs	Must be at least 30%
Labor costs compared to material cost	Should be 50%–50%
Number of maintenance employees compared to plant population	15%–25%
Total maintenance costs compared to the cost of goods	Varies, 4%–8%
Maintenance costs per unit produced	Site specific
Maintenance costs per square foot maintained	Varies by site
Production loss caused by maintenance	Varies by site
Estimated replacement value per maintenance worker	Varies by site
Union attitudes	Varies by site
Management support	Varies by site
Corporate environment	Varies by site

Once these base indicators (or the utilization of other indicators) have been established, the plan to develop the maintenance organization can be formulated. The present benchmark should be documented and the projected improvements established. (For further information concerning maintenance and organizational evaluations, see *World Class Maintenance Management* and *Total Productive Maintenance—An American Approach*, both published by Industrial Press, Inc.)

Once the analysis is conducted and the organization is benchmarked, the goals should be established for each of the analysis points that are to be monitored.

WARNING

Many organizations are becoming obsessed with establishing and monitoring benchmarks. Some go to great efforts to set "standards" for their market or industry. However, benchmarks are only useful for inspiration. If you set some industry "standard" as a goal for your program, you will never be competitive. If your competitors are already at these standards, where will they be 3 to 5 years from now? When you have achieved your competitors' current status, their continuous improvement programs will still make them 3 to 5 years ahead.

A Second Warning: Many organizations concentrate on an extensive analysis, and it takes so long that they never really get their programs started. It has been said that conducting extensive analysis is the best method an organization can use to postpone doing anything really useful!

Setting Goals

Examining the above indicators should highlight the five goals for the TPM program.

1. Ensure equipment capacity (or facility availability).
2. Utilize a program of maintenance for the entire life of the equipment.
3. Require support from all departments involved with the use of the equipment or facility.
4. Solicit and utilize input from employees at all levels within the organization.
5. Utilize consolidated teams for continuous improvement.

The indicators or analysis points that are utilized should be focused on these goals. For example, ensuring equipment capacity can be highlighted by examining the percentage of maintenance-related delays. The life cycle program for the equipment can be monitored by the costs of the preventive maintenance program. The indicators should be chosen and used carefully, since they will form the information base for determining when your goals have been achieved.

Building the plan of how to get from the current condition to the goals varies from company to company, depending on the present condition of the organization and the resources available to work on the project. The program should be viewed by all employees — from top management to the line personnel — as essential to the survival of the company. When the program has sufficient priority, only then will it be successful.

Education for a TPM Program

The training for a TPM program is divided into two categories: philosophy and technical. The philosophical training is critical so everyone understands what the program is supposed to accomplish. The technical training is to provide the employees with the information on how to perform the tasks that are assigned.

The philosophical training is presented in the form of bottom line dollars. It takes time and effort to convert breakdown information into quantified dollar amounts. Once organizations do this, they find that downtime on equipment is much more costly than anyone ever imagined. Some companies find that equipment downtime may be worth $500, $1,000, $10,000, or more per hour. Others will find that they have stepped costs when the equipment is down. For example, for the first hour, the cost may be very small. But, when the equipment is down over an hour, because of the JIT product flow, it shuts other equipment down. After 4 hours or so, the entire process or plant may be down. Then the downtime is very expensive. There are still companies who say "there is no real cost involved if the equipment is down; we will just make it up on another shift." This is still the old "fat cat" type of thinking. If the production schedules have that much "fat" in them, then the operation is not competitive with the world market. If they just work overtime to make it up, there is still the added energy costs to run the equipment, the premium for the employees to be on the job, etc. It goes back to the old saying "there is no such thing as a free lunch."

The philosophical training also must include the theory behind TPM. This can be presented in one or two short sessions to the various groups. This is to ensure that all employees are working toward the same goals at the same time. Without this background, there will be opposition to the program. The true goal in the program — competitiveness — must be clearly established. All other subgoals are just methods to achieve the competitive edge that TPM gives a company.

The technical education involves the basic operational theory behind the equipment: the mechanical, electrical, and fluid power systems. Information to be covered would include the theory of operations, how to maintain the components, and how to inspect the components. The technical training will be on two levels. The operators will need to know the basics to safely perform their job duties. The maintenance technicians (or team members) will need to know more advanced information to allow them to diagnose and troubleshoot chronic and technical problems.

The operators should never be trained to be maintenance technicians. There is no need for this amount of information, since they are to be the first line of defense when equipment problems occur. A common problem for a TPM program is for the various groups to lose focus, and then the entire program suffers. By providing each group with the information it requires, the program will be successful and cost effective.

2 Preventive Maintenance

Preventive maintenance is a philosophy, not just a term. It is rapidly replacing common maintenance practices in competitive industry today. It's not a change that's coming easy. Craftsmen and supervisors don't change their way of thinking and their habits quickly. However, preventive maintenance must become part of the maintenance organization's very thinking process if it's to be effective. Without this type of program ingrained into an organization, maintenance costs become excessive. A typical cost-versus-delay curve is illustrated in Figure 2-1. To cover this cost, the price of the manufactured product must be raised. Such a rise in price will prevent the manufacturer from being competitive in the marketplace.

What is Preventive Maintenance?

Preventive maintenance (PM) is action taken to keep an item which is in operation in an operating condition by means of inspection, detection, and prevention of failures. When we look into everyday practice, we see many examples of preventive maintenance. You don't undercoat your car unless you're trying to prevent rust.

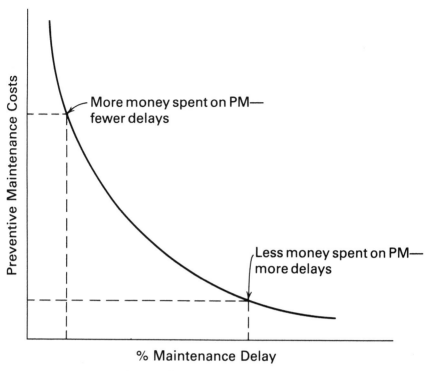

Figure 2-1. Cost-versus-delay curve.

You don't wait until the oil light comes on in your car before you change the oil.

Why Do We Need Preventive Maintenance?

Why have preventive maintenance? Here is a place where Murphy's law finds an application. Equipment failure usually occurs at the worst possible time. For example, the production department may be in the middle of a rush order that is very critical to the customer. This is the time the equipment will break down and cause an interruption in production. Depending on the severity of the failure, the equipment may be down from a few hours to a few days. This delay can prove very costly in the competitive marketplace that all industry finds itself in today. How much better it would have been if the maintenance department had been able to detect a problem in the equipment, arranged with the production department a convenient time, and made the necessary repairs before a failure occurred. This is the ideal situation, and can only be achieved after the program has been in use for a period of time, and after proper training has been given to those performing the inspections.

A second reason for preventive maintenance is safety. Proper inspections can detect unsafe conditions in time to prevent an accident, which might cause damage to the equipment or injure operating personnel.

A third reason is reduced repair costs. When a failure occurs, it usually destroys equipment that is associated with the defective component. If the defective component is changed before the failure occurs, the related equipment will not be damaged; thus, repair costs will be reduced. With the price of replacement parts escalating at today's rate, this cost savings can be quite substantial.

Types of Maintenance

Maintenance types can be broken down into four main classifications:

1. breakdown
2. corrective
3. renovative

4. preventive
   (a) monitored
   (b) scheduled.

Breakdown Maintenance

Breakdown maintenance refers to equipment breakdowns occurring before any maintenance has been performed. This is also referred to as unscheduled maintenance.

Breakdown maintenance is practiced in industry today. Sometimes it's necessary, especially on low cost components where the equipment is of an auxiliary nature and not directly related to production. If the component that is down doesn't cause an interruption in production, repairs can be effected without interruption of the flow of materials. If the equipment doesn't endanger safety, preventive maintenance may not be practiced. The cost of performing preventive maintenance should be considered in all applications before it is initiated to ensure that the cost of the program does not exceed the equipment costs. When considering larger, more expensive components, care must be taken to figure in cost of probable damage to related components, along with the cost of the lost production time and wasted man hours.

Corrective Maintenance

Corrective maintenance is merely repair work, which may be performed on a scheduled basis or during inspection times. It's usually used in correcting a defect before the breakdown of the component occurs. This type of maintenance is usually performed in response to a preventive maintenance inspection. The effect that corrective maintenance has on equipment availability is shown in Figure 2-2. As the amount of corrective maintenance performed is reduced by PM inspections or condition monitoring maintenance, equipment availability increases. Care must be taken to avoid either extreme; otherwise it's not cost effective. Somewhere along the line is the most economical situation.

Renovative Maintenance

Renovative maintenance is performed when the equipment can be taken off line for an extended period of time. This scheduled outage or shutdown of the equipment should be scheduled with production to prevent a bottleneck around the equipment. This type of maintenance usually refers to major modification, redesigns, or installation of some technological advancements. The cost effectiveness of this type of maintenance is shown in Figure 2-3. In the beginning, it's cheaper to operate the equipment and replace parts as they become defective. However, when the cost to overhaul becomes greater than the cost to replace, then it's time to schedule a period of renovative maintenance. After performing this type of extensive maintenance on a piece of equipment, it's usually restored to a condition close to new. The breakdown rate should fall off dramatically, until the next period of renovation is needed.

Scheduled and Monitored Preventive Maintenance

Preventive maintenance can be divided into two basic categories: scheduled and monitored.

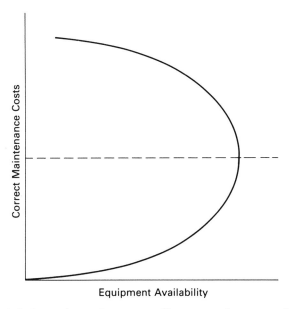

Figure 2-2. Corrective maintenance effect on equipment availability.

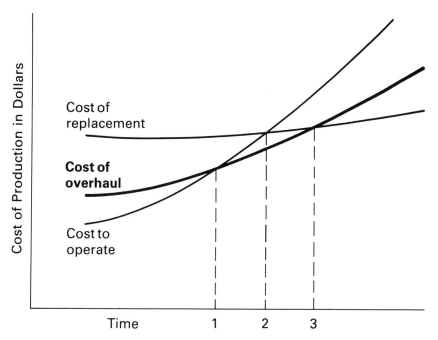

Figure 2-3. Cost effectiveness of renovative maintenance.

Scheduled maintenance is the process of periodic overhauls or service of the equipment. This may be anything from a lubrication routine to a time frame for replacement of component parts on a piece of equipment; the point is that maintenance is set to a time period, much the same as service to a car is tied to mileage. The time may be in *hours of operation, number of shifts of operation,* or a given *service period in days, weeks, or months.* Once the schedule has been set, periodic checks should be run from time to time to ensure that the times set up are correct, and that the equipment is not being inspected too frequently or not often enough.

Monitored maintenance uses sophisticated testing equipment to help predict when the equipment components will fail. This testing equipment can even be interfaced with a microprocessor to chart equipment wear rates for even better estimations of equipment condition. Such a system allows logical decisions to be made as to replacement of worn parts without causing lost production hours, for the equipment can be changed on scheduled repair turns. This helps take the guess work out of component replacement. Figure 2-4 illustrates how a control limit is set defining the amount of wear that is acceptable. Once this point is exceeded, the component should be changed. If it isn't replaced, then the failure area will be reached resulting in a breakdown. If changed when the control limit is reached, it can be scheduled so as not to interfere with production.

As we progress through the preventive maintenance program, notice that all of the previous types of maintenance have their place in the organization. The cost of the equipment, lost production, wasted man hours, and repair times will be weighed against the cost of preventive maintenance to see what equipment needs it and what doesn't.

Any properly designed and operated preventive maintenance program will more than pay for itself. The initial setup and operation will increase the overall costs; but after a period of time, the maintenance cost will drop off below the original level (see Figure 2-5).

How to Sell Management on a PM Program

If your plant doesn't have a preventive maintenance program, what's the best way to persuade plant management that a program is necessary? Let's begin by compiling the following records.

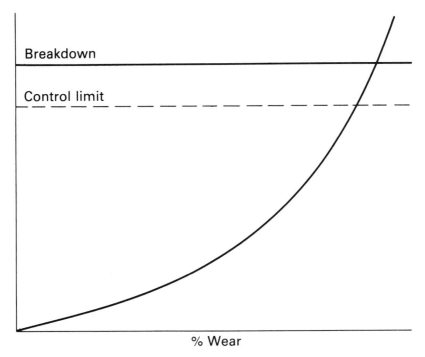

Figure 2-4. Control limit versus wear.

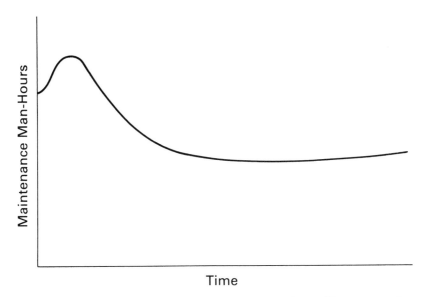

Figure 2-5. Initial set-up costs are offset by savings achieved by a preventive maintenance program.

1. List total breakdown repairs (include repair labor, materials, overtime, etc.).
2. List idle production time (include production labor, cost of equipment delays, defective products produced).
3. List overhead and any equipment damage resulting from the breakdown.

After compiling these figures for a time period (a year should be minimum), figure the cost for the same repairs if they could have been scheduled at a noncritical time period. This would include the time saved by having the necessary parts when repairs began, the costs saved by having the production and maintenance force scheduled for the repair, and the cost savings from the prevention of damage to related components. The difference between the two cost figures will be the savings that the preventive maintenance program will be able to attain. This method is usually sufficient to point to the value of the preventive maintenance program.

Where Do You Start?

Unless an installation is small, you can't start a plant-wide preventive maintenance program in all areas at the same time. Where should it be started? It depends on each plant's circumstances. A program can be started in one department, and when finished, a second can be selected and set up. This process can continue until the entire plant is completed.

Another method is to select a certain type of equipment and design a preventive maintenance program for it. Then take that program and apply it to all similar equipment plant-wide. Next, choose another type of equipment and set up the preventive maintenance system for it, and so forth until all types of equipment are finished.

If the program must produce immediate results, it's best to pick the most obvious trouble area and set up a program for it. By concentrating all efforts in one area, the results will become noticeable very quickly. This will produce the needed enthusiasm and cooperation necessary to complete the program plant-wide. By having a good case, it will be easy to sell the plant management on the advantages of preventive maintenance.

Cost of PM Versus Cost of the Equipment

One of the most important considerations in a preventive maintenance program is the cost of the program versus the cost of the
equipment. It would be a waste of time and money to spend several
hundred dollars to inspect and maintain a part that only costs $25
and would not interrupt production in case of failure. Some plants
set limits on equipment inspections, holding them to a certain cost;
for example, inspect above $100, don't inspect below $100. The limit
would have to be set by each plant for its particular equipment.
Some guides to follow are as follows (also see Figure 2-6).

1. Inspect any item that will cause a major shutdown, lower quality,
   or costly damage to related components, or produce a safety hazard to employees.
2. Plant fixtures such as lighting, flooring, or ceilings that would interfere with producing a quality product, or would produce poor
   working conditions, should be included in a preventive maintenance inspection.

Items that become questionable for inclusion in a preventive
maintenance inspection program might include the following.

1. Equipment that has a backup or standby system. In case of a
   breakdown, the secondary system can be operated while repairs
   are made to the primary system.
2. Equipment that costs no more to repair than it does to perform
   preventive maintenance. If the cost to take the equipment apart to
   repair it is lower than or the same as the cost of removing a defective item found on an inspection preventive maintenance, then
   it is highly questionable to inspect it.
3. Equipment that lasts long enough to meet minimum life requirements without preventive maintenance should not be included in
   the preventive maintenance inspection program.

Inspection Intervals

After the decision is made as to which equipment will receive the
preventive maintenance inspections and be included in the program,

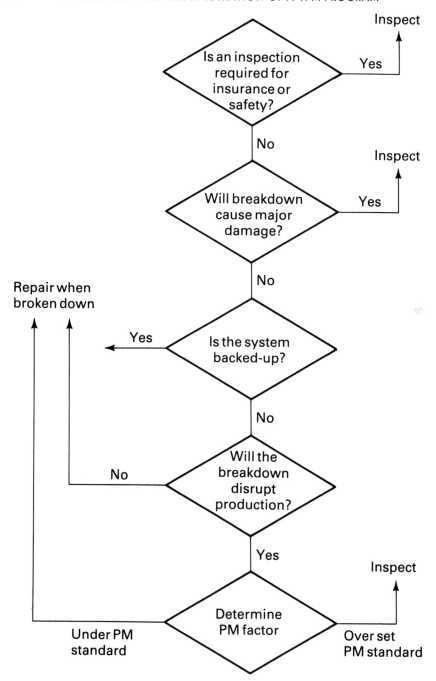

Figure 2-6. Deciding on what to inspect.

the decision must be made on how the inspection intervals should be spaced.

This time interval will also vary with the age of the equipment. The typical "bath tub" curve (Figure 2-7) has an application in figuring the service versus the age of equipment. As the equipment is started up, it has a break-in period when the failure rate is high. As the equipment gets a little older and the bugs are ironed out, the failure rate becomes fairly consistent. As the equipment ages, the individual components begin to wear out and the failure rate begins to go up again. It will become important to adjust the preventive maintenance times according to this curve. Usually as the equipment enters the last part of the curve, an overhaul or renovative maintenance is performed, resetting the equipment to an earlier part of the curve.

Preventive maintenance timing falls into three classes (Figure 2-8):

1. too little
2. just right (which is rare)
3. too much.

The too-little or not-frequent-enough service (#1 in the figure)

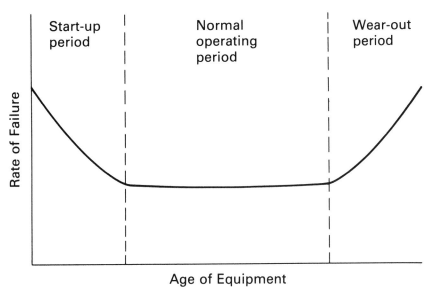

Figure 2-7. Rate of failure versus age of equipment.

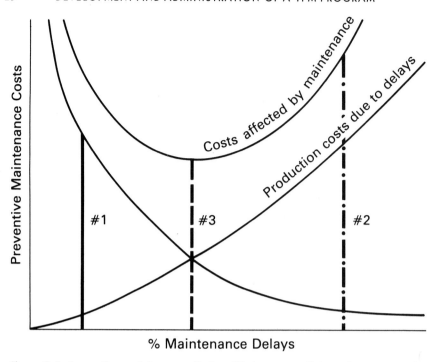

Figure 2-8. Preventive maintenance timing: #1 shows too little maintenance, #2 shows too frequent maintenance, and #3 shows the correct amount of maintenance.

will be evident by the excessive number of breakdowns that occur. The equipment fails before it receives the proper service. The other extreme is too much or too-frequent service (#2 in the figure). This is a waste of manpower and a waste of components, which are changed before they wear out. This adds unnecessary expense to the preventive maintenance program. The question is how to determine if there is too much preventive maintenance.

The program should be evaluated to see when the failures are occurring. If there are no failures, chances are you are performing too much preventive maintenance. It might reduce expenses to lengthen the service times. About 20% of the equipment should fail before service if the preventive maintenance times are set correctly (#3 in the figure). If the failure rate is less, try lengthening the times to reduce costs. If the failure rate is higher, try reducing the time between services to prevent the breakdowns. Accurate records are necessary to properly set the time schedule for preventive maintenance.

While this method is to be applied as a rule of thumb, there are many more complicated methods available to maintenance engineers using statistical analysis and probability models that can give a more exact time frame, should this be necessary. One thing to keep in mind is that either extreme will be costly. Whatever action is necessary to achieve the lowest cost program with the proper schedule of PM activities must be taken. If the simplified method will serve the purpose, then use it. If not, you may want to consult an engineering firm dealing with maintenance in order to establish the correct frequency. Whichever method you use should give the necessary results — lower cost maintenance, which is the bottom line.

Inspection

Inspection is the process of examining equipment to:

1. make sure it's performing as designed
2. evaluate all components for potential problems
3. identify any component that may cause a breakdown, and estimate the time until failure.

The inspections should be forwarded to a scheduler to ensure that repairs can be made at a time that will not interfere with the production department.

There are basically three types of inspections:

- those inspections required by law
- those made on equipment with no backup
- those made on equipment while it is broken down.

A formula that may be used to determine if it is cost effective to inspect equipment is:

$$PM = \frac{D(A + B + C)}{(E \times F)}$$

where

PM = inspection factor

D = number of breakdowns/year

A = cost of breakdown repairs

B   = cost of lost production

C   = cost of repairing other equipment involved in the break-
        down

E   = cost of PM activity (average)

F   = number of PM cycles/year.

Deciding Which Equipment Should Receive PM

When deciding which equipment will receive PM, you should start with those items having the highest PM factor and work toward the lowest. The process should be carried out until the cost of the program is equal to the amount budgeted.

The next consideration should be given to the time allotted to the performing of the preventive maintenance inspections and repair jobs. Again, it's necessary to keep accurate records. The time allotted to each job and the actual time it took to perform the work should be recorded. After the information has been gathered over a time period, a check should be made. The suggested way is to compile the average of the figures of the estimated times. This should not be done until the job has been assigned numerous times and a good cross section of times is available. It may be as high as 100 times, and should not be less than 50 times for an accurate analysis. Then the average should be taken of the actual time it took to perform the work. If the difference is within a range of ±20%, the estimated time is accurate enough for most applications. If it's outside this range, then some adjustments need to be made in the estimated time to correct for the difference. This method will allow for the most optimum utilization of all personnel involved.

Inspection Reports

Reports need to include several items. The first is the inspection frequency. How often is the equipment to be inspected — daily, weekly, monthly, semiannually, or annually? This cannot be established without serious consideration given to the equipment and its individual service requirements. Some sources that may be consulted include:

1. manufacturer's service recommendations
2. equipment installer's recommendations

3. maintenance craftsmen's recommendations

4. spare parts suppliers.

Once the frequency has been decided upon, the forms should be made up. The inspection reports should be clear and concise, easy to read, roomy, and specific. Codes utilizing check marks or x's should be used for simplification of the inspection process.

Card System for PM

With the card system, each piece of equipment is assigned six separate cards, as follows.

1. A record of equipment spares and replacement parts (Figure 2-9); this should include print number and part stock number.
2. A preventive maintenance card, showing the PM assignments and their frequency (Figure 2-10).
3. & 4. A history record (Figure 2-11) including all breakdowns, work performed, and parts used. [This should also include a report on why the breakdown occurred (Figure 2-12).]
5. A preventive maintenance inspection card (Figure 2-13), which would include the item to be inspected and its condition. This form can be passed on to the scheduler to see that any defective items are corrected at the earliest opportunity.
6. A PM checklist (Figure 2-14). This is to ensure that all items are inspected at the correct frequency. This will serve as a reminder so that none is missed.

These cards may be kept in separate files and continually updated as necessary.

A daily PM schedule can be made by listing all of the daily PM checks and services to be made on each piece of equipment. Once this list is compiled, it can be used as the daily PM service sheet. A weekly, monthly, semiannual, and annual list can be made the same way. One point to remember: it's a waste of time to make these lists if they're not going to be used. For the system to be effective, the inspections and services must be carried out according to schedule. Some flexibility must be used, but don't start making all of your daily checks and services weekly or the breakdown rate will increase tremendously. With all inspection forms and worksheets, the KISS formula should be used (Keep It Simple, Stupid).

Equipment Spares			
Replacement Part	Blueprint Number	Stock Number	Location
⋮	⋮	⋮	⋮

Figure 2-9.

Preventive Maintenance Schedule
Equipment _____ Equipment Number _____ Location _____
Daily
Weekly
Monthly
Yearly

Figure 2-10.

Equipment History			
Equipment _____ Equipment Number _____			
Date	Problem and Solution	Labor	Parts Used
⋮	⋮	⋮	⋮

Figure 2-11.

Failure Analysis Report
Date _____   Equipment Number _____   Equipment _____
Failure
Cause of Failure
Recommended Solution
Results of Solution
Date of Observation

Figure 2-12.

Preventive Maintenance Inspection									
Check the column that indicates the condition of the unit or what problem exists.	O.K.	Needs lubrication	Requires adjustment	Requires replacement	Requires cleaning	Excessive vibration	Excessive heat	Loose	See Additional
⋮	⋮	⋮	⋮	⋮	⋮	⋮	⋮	⋮	⋮
Additional Comments									

Figure 2-13.

Preventive Maintenance Check List				
Month _____				
PM Assignment	To be performed	Date	Employee	Comments
⋮	⋮	⋮	⋮	⋮

Figure 2-14.

Planning PM

The four basic preventive maintenance functions are:

1. lubrication
2. inspection
3. overhaul
4. part replacement.

To make proper use of maintenance manpower, 70% of the time should be scheduled. If more than 30% is used in breakdown or emergency maintenance requests, then productivity of the maintenance staff suffers. The emphasis placed on PM will pay off with greater benefits: more machine production and less delay time. Let's look at the four PM activities in detail.

1. **Lubrication:** This includes checking lubrication levels, filling to the proper levels, and changing lubricants.
2. **Inspections:** This may include visual observations and testing with sophisticated equipment for the purpose of determining where a component is in its life cycle.
3. **Overhaul:** This includes scheduling equipment down so that major repair work can be performed on it. It's the same as renovative maintenance, which was discussed earlier.
4. **Part Replacement:** This is replacing a defective component without a major overhaul. It keeps the equipment out of service for a relatively short period of time.

If you make your schedules while keeping these four areas in mind, planning and scheduling PM becomes easier to coordinate with the production department. This will enable the plant to run at maximum efficiency. The main thing is to make everyone a part of the program, from the craftsman to the production worker. This will kindle a spirit of cooperation that will ensure the success of the program.

3 Predictive Maintenance

We all know about the three R's: reading, 'riting, and 'rithmetic. Well, there are also three R's of maintenance: routine, renewal, and repair.

If you ignore the first two, the third is inevitable. Routine involves three things: inspection, adjustment, and lubrication (see Figure 3-1). Inspection is the main part of predictive maintenance.

Inspection is a means of predicting a need for future work. It may consist of visual inspections or the use of condition-monitoring equipment to check all wearing parts to avoid a breakdown. Remember — dirty filters, leaky seals, excessive noise, vibration, or heat can only be seen or heard if the inspections are made. It will do no good to make up the inspection sheets or buy the inspection equipment if the inspections are never carried out.

Equipment Wear Records

Predictive maintenance utilizes the above methods to predict failures. An essential part of any predictive maintenance program is the equipment records. The wear of the equipment must be plotted

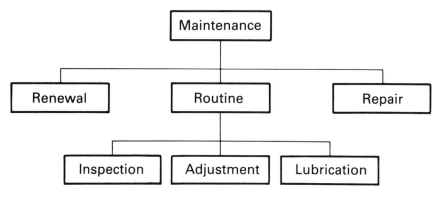

Figure 3-1. Relationships of maintenance functions.

against time. As the inspections show the wear progressing, a graph is made. When the component fails, it's noted on the chart (see Figure 3-2). When it's replaced, a new chart is started. This time the component is monitored and is replaced as it approaches the life of the last component, provided it's wearing at approximately the same rate. This prevents a breakdown and allows the maintenance staff the flexibility to schedule the repair without interrupting production — the advantage of predictive maintenance.

The process is not as easy or simple as outlined in the above paragraph. However, with a little effort, the system can be instituted to complement any preventive maintenance program.

Establishment of Standards

One of the most complicated procedures is the establishment of the standards (Figure 3-3). If the equipment is replaced too frequently, it increases costs. If it's not replaced frequently enough, breakdowns occur. The best method is to research equipment life. Note the average time for replacement of a component and try that time period. If it works, then use that time period. If it doesn't, then shorten it slightly, until your replacements come before your breakdowns. From time to time it may be necessary to let a component run until breakdown, just to ensure that your time periods are approximately correct.

What type of equipment should predictive maintenance be practiced on? Since cost is involved, PM should be used only on compo-

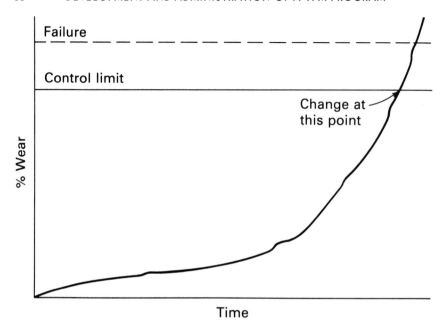

Figure 3-2. Control limit graph.

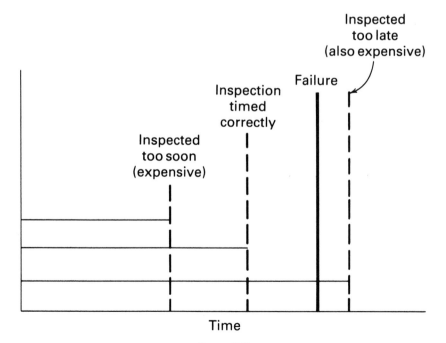

Figure 3-3.

nents whose breakdown will lead to long periods of downtime. This is a matter the maintenance supervisor should discuss with the production supervisor. This will ensure the maximum efficiency of the program.

Who should perform the inspections? To evaluate the life of a component, it takes a qualified individual. Suitable training and inspection guidelines are necessary. Some of this information is provided in the second half of this book. Inspecting equipment also requires the use of suitable instruments capable of providing the necessary accurate readings. For example, some engineers recommend that for every $100 value of a piece of equipment, there should be $1 spent on condition-monitoring equipment. While this seems expensive, it actually reduces costs because decisions are made on factual information, not guesses.

Equipment Used in Predictive Maintenance

What kind of equipment is usually used? The cornerstone of predictive maintenance is **vibration analysis** (see Figure 3-4). This is a measure of the amount of vibration of certain components in a drive system. When the equipment is new and correctly installed, vibration is usually very low. As the equipment begins to wear out, the vibration increases. If accurate readings are taken at periodic intervals, the increase in vibration can be plotted. When the vibration reaches a certain level, indicating considerable wear, the component can be changed before a failure occurs. (More detailed information on this subject will be considered in Chapter 4.)

While this may seem to be an oversimplification, that's the basic principle behind vibration analysis. This method, coupled with visual inspections, provides a low cost method of predictive maintenance. If an inspector has the proper forms and a portable vibration meter, effective results can be achieved.

Inspections in Unsafe Locations

The next problem may be how these inspections can be carried out in locations where it would be unsafe to test while the equipment is operating. This can be solved by using **vibration transduc-**

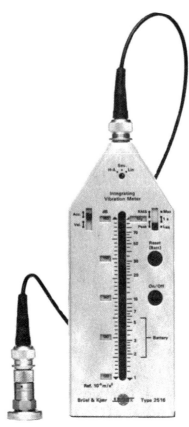

Figure 3-4. Vibration meter. (Courtesy of Brüel and Kjaer Precision Instruments.)

ers that can be mounted right on the equipment (see Figure 3-5). This style of transducer can be permanently mounted on the equipment. The signal can be fed into a control room where it can be attached to a meter. The meter may be able to accept feeds from several transducers (Figure 3-6). This will allow more economical monitoring of the equipment. The readings can then be recorded and charted to give an indication of the condition of the equipment.

The next step up in sophistication would be the attachment of a **graphics chart** to the monitor to give a printout of the condition of the equipment (Figure 3-7). The charts can then be collected and sent to a central location for analysis. This is a good method when the number of trained personnel is limited.

The ultimate in sophistication is a **computer feed** that will moni-

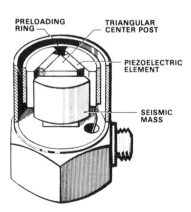

PRELOADING RING — TRIANGULAR CENTER POST — PIEZOELECTRIC ELEMENT — SEISMIC MASS

Figure 3-5. Vibration transducer. (Courtesy of Brüel and Kjaer Precision Instruments.)

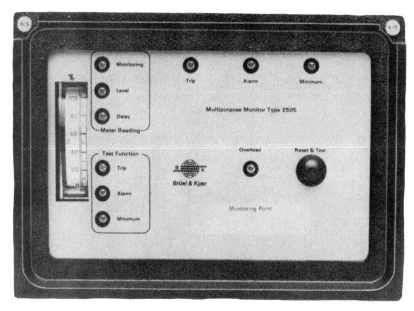

Figure 3-6. Monitor panel. (Courtesy of Brüel and Kjaer Precision Instruments.)

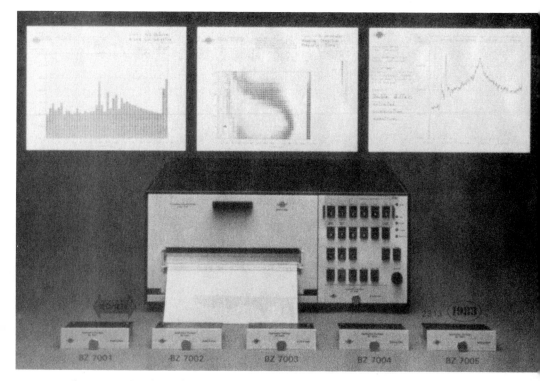

Figure 3-7. Graphics charts. (Courtesy of Brüel and Kjaer Precision Instruments.)

tor the vibration level. The computer can give a chart of the progress of the vibration, and even sound alarms when the vibration exceeds certain levels. Of course, the funds available, coupled with the critical nature of the equipment, will determine the amount of equipment necessary for each installation.

Other Forms of Equipment Monitoring

In addition to vibration analysis, there are other forms of equipment monitoring or nondestructive testing that are available for determining the position a particular component is at in its life cycle. Some of the other methods are: **acoustical emission** monitoring, i.e., the monitoring of sonic waves which can come from cracks that are beginning or progressing, or are from leaking air or fluid lines; and **oil analysis**, which analyzes the contaminants in oil to determine wear levels.

Predictive maintenance is the ultimate in preventive maintenance inspections. By proper use, it can be determined when a breakdown will occur, and will thus eliminate the unnecessary breakdowns and disruptions to the production departments. This method will be much more accurate than the visual inspections used in preventive maintenance programs.

4. Nondestructive Testing

In industry today, we need to know the amount of deterioration in a piece of equipment, or the condition of a finished product, without destroying or dismantling the product. These types of data will ensure product reliability without causing long delays on production equipment, or allowing equipment to deteriorate to an excessive degree before some type of maintenance activity has been started. There are five main types of nondestructive testing used in preventive maintenance activities:

1. liquid penetrant
2. magnetic particle
3. ultrasound
4. vibration analysis
5. oil analysis.

These are not the only forms of nondestructive testing in use, but they *are* the most common and are therefore the ones whose application to maintenance will be considered at this time.

Liquid Penetrant

This method of nondestructive testing and inspection is used to find surface cracks or subsurface defects that have an opening to the surface. Liquid penetrant testing is performed in the following steps.

1. Cleaning of the surface to be tested.
2. Application of the dye.
3. Removing the dye from the surface after the proper penetration time has expired.
4. Application of the developer.
5. Inspection of the results.

The process begins by cleaning the surface to be tested. This is very important because dirt or contaminants may fill the crack or defect and not allow the dye to penetrate (Figure 4-1). The surface to be tested must be free of any paint or coating material. Rust, scale, oil, or grease must also be completely removed. If these are not removed, then the test results won't be dependable. Once the surface is clean and dry, the test may proceed.

The next step is the application of the dye. There are several types of dyes available. They can be divided into two main categories. The first is by the type of light that the dye is visible under — ultraviolet or white light. The dye that's visible under ultraviolet light requires the use of a black light (or ultraviolet light) for the dye to be visible. The other dye is visible under bright white light.

Another method of division is the means by which the dye is removed from the tested surface — water washable, solvent remov-

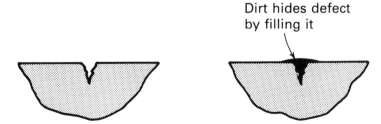

Figure 4-1. The surface to be tested should be cleaned because dirt fills in the defect and hides it.

able, or postemulsification. If the dyes can be removed by washing with water, they are water soluble. If they must be removed by the application of a special solvent, then they fall into the second class. If they must have a special emulsifier applied and then be washed with water, they fall into the third class.

Once the dye has been applied, it must be allowed to set for a period of time so that the dye will penetrate into the cracks or defects that are present. Capillary action accomplishes this step. This time varies depending on the dye used and the material being tested, but runs between five minutes to over an hour.

After the time has passed, the dye must be removed from the surface of the material. Depending on the type of dye used, the removal method varies from flushing with water to the application and removal of special solvents or emulsifiers. This removes the surface dye but leaves the dye that has penetrated into the cracks or surface defects.

The developer is now applied [Figure 4-2(c)]. The application method depends on whether you use wet or dry developer. The developer draws out the dye that has penetrated into the crack or defect. It appears above the crack and spreads out to indicate the location of the defect.

The last step is the inspection of the result. This may require only ordinary light if visible dye was used, or may require the use of a black light if a fluorescent dye was used. The important thing is to be able to read the results to come up with the proper conclusion. This can only be accomplished by proper training and experience.

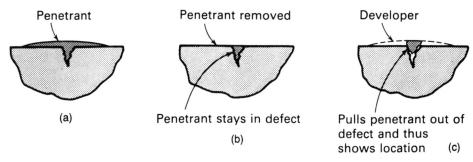

Figure 4-2. Liquid penetrant method of testing deterioration of equipment. (a) Application of the dye, (b) removal of penetrant, (c) application of the developer.

Magnetic Particle

This nondestructive testing method uses the principle of magnetic fields to indicate defects in materials that can be strongly magnetized. It's not effective on material that cannot be magnetized, for example, brass, copper, aluminum.

This testing process consists of four basic steps.

1. Cleaning the surface to be tested.
2. Developing a magnetic field in the object being tested.
3. Applying magnetic particles to the object.
4. Examining the pattern of the particles for indication of discontinuities.

To begin the test, it's important to clean the area being tested, otherwise dirt and grease may interfere with the accurate reading of the results. Rust and scale should be removed because they're usually magnetic and will contaminate the magnetic particles that are applied.

In developing a magnetic field in the object being tested, it's important to realize that there are several different types of magnetic fields that can be developed. In a ring-shaped object the magnetic field takes one shape (Figure 4-3); in a bar shape it takes another (Figure 4-4); in a rod-shaped object it takes another (Figure 4-5); and in testing a plate the field takes another shape (Figure 4-6). The type of equipment used will determine the correct attachments and positioning of the fields. It's best to consult the instructions for the particular equipment and type of material being tested.

The third step is the application of the magnetic particles. These are usually divided into two classes — **wet** and **dry**.

In the wet method, the particles are suspended in a suitable carrier. Also, the particles may be fluorescent for easier readings under a blacklight. The wet particles are best suited for finding very small and fine defects.

The dry particles are best suited for use on rough surfaces and for finding some subsurface defects. These particles usually come in the form of a very fine powder. They may come in several colors, including fluorescent. The choice of one color over another should be made on the basis of which one will give the best contrast against

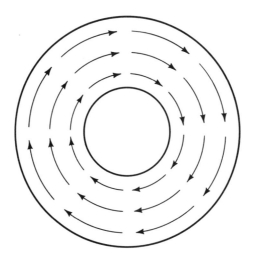

Figure 4-3. Ring-shaped magnetic field.

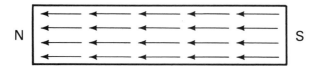

Figure 4-4. Bar-shaped magnetic field.

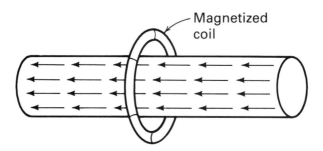

Figure 4-5. Rod-shaped magnetic field.

the object being tested. Application of the powder to the part being tested must be made evenly and gently. The particle must not be forcibly applied because this will interfere with the attraction of the particle by the magnetic field. The powder can be applied gently with a small spray bulb or a small shaker. For larger or multiple tests, a small, special powder blower may be more practical to use.

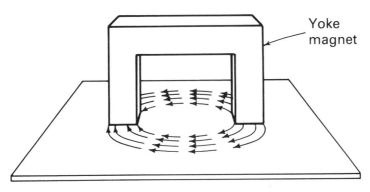

Figure 4-6. Yoke-shaped field for plate test.

Examination of the pattern of the magnetic particle is the next step. The defect will show up as a disruption of the smooth flow of the magnetic field. Again, the inspection will depend on the shape of the magnetic field. If the object being tested is a ring, the defect should show up as pictured in Figure 4-7. If the object is a bar, it shows up as pictured in Figure 4-8. If the object is a cylindrical shape, the defect is as shown in Figure 4-9. A plate shape magnetized by a horse shoe or yoke magnet looks like that shown in Figure 4-10.

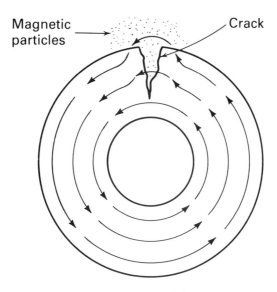

Magnetic particles   Crack

Figure 4-7. Ring defect.

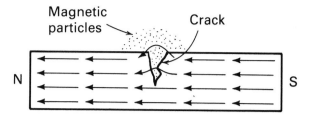

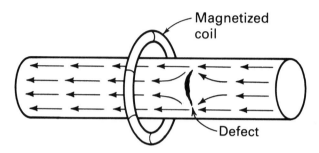

Figure 4-8. Bar-shaped defect.

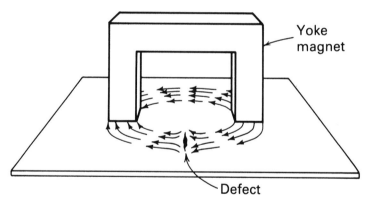

Figure 4-9. Cylindrical or rod-shaped defect.

Figure 4-10. Defect shows a pattern running perpendicular to the lines of force.

One last consideration in concluding a magnetic particle test is the effect the magnetic field has on the object being tested. Some objects have a tendency to hold the magnetic field after the test. This may interfere in some further process or installation the object may be involved in. If this is the case, it may be necessary to demagnetize the object after the test. The procedure for the demagnetization depends on the object, the amount of residual magnetism, and the method used to magnetize. It's best to consult the manufacturer's recommendation for the correct method of demagnetization.

Ultrasound (Ultrasonic) Testing

Ultrasonic testing is the use of high-frequency sound waves transmitted into or through an object. The results of the test are read on a meter and will give two results: the thickness of the object, and any subsurface flaws in the material.

Ultrasonic testing is becoming more and more common as a method of quality control, especially in heavier industries. Any industry that manufactures heavy, thick material can now inspect the internal structure without damaging the product.

Preventive maintenance can also benefit from ultrasonic testing because the inspector can detect subsurface defects from fatigue, corrosion, or initial manufacture. The process is the same whether for quality control or for preventive maintenance.

The process is as follows (Figure 4-11).

1. The transmitting unit is attached to the tested part.
2. The receiver is placed on the other side of the part or beside the transmitter, depending on the type of test and the type of equipment being used.
3. The transmitter is activated.
4. If there's a flaw in the material, the signal or part of the signal is

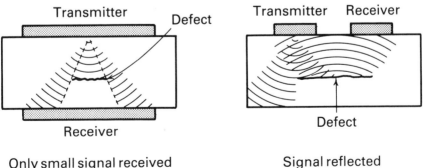

Only small signal received
due to blockage by defect

Signal reflected
by defect

Figure 4-11. Ultrasonic inspections.

reflected back to the receiver. If part of the signal is reflected back, it indicates a subsurface defect.

Depending on the type of equipment, a visual picture of the type of defect may be available. The systems are even so sophisticated that high-speed inspection of a large volume of symmetrically shaped objects is available. For preventive maintenance applications, this method can spot fatigue cracks in equipment before they become severe enough to cause equipment failure. The crack can be monitored and the equipment scheduled for repair before a catastrophic failure occurs.

Vibration Analysis

Webster defines "vibrate" as a rapid back-and-forth movement. All machinery vibrates no matter how carefully it has been constructed and installed, and no matter how close the tolerances it was manufactured to. Any change in the condition of a piece of equipment is usually accompanied by an increase in the amount of vibration. By monitoring the amount of vibration of a certain piece of equipment, its condition can be determined. Vibration analysis takes this one step further. It analyzes the cause of vibration instead of just determining the level of vibration.

Vibration has two main characteristics: frequency and amplitude. **Frequency** is the number of times that a part oscillates through a cycle in one second. This is usually expressed in hertz. The other measure is cycles per minute (CPM). This measure is usually reserved for equipment of a slower speed. **Amplitude,** which is the measure of the severity of the vibration, can be measured in three ways (Figure 4-12):

1. displacement
2. velocity
3. acceleration.

Displacement is the actual amount of movement that takes place. Usually it's measured in peak-to-peak displacement, which is the total movement that takes place. It's usually measured in thousandths of an inch.

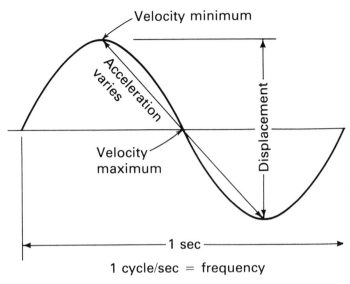

Figure 4-12. Measurement of amplitude.

Velocity is the speed at which the displacement takes place. It can be compared to traveling at different rates of speed in your automobile, for example, 50 mph, 25 mph, 45 mph. This velocity is usually expressed in inches/second, since it's usually not traveling too great a distance.

Acceleration is the time rate of change in velocity. When your car is at rest and you speed up to 60 mph in 5 seconds, you are accelerating 12 mph per second. So when a rotating object is at the peak and reversing direction, it increases to peak velocity in a short period of time; this is the maximum acceleration.

The three parameters making up amplitude (displacement, velocity, and acceleration) are usually the measurements taken to determine vibration. Displacement is usually used to measure low- or constant-speed machinery. Velocity is used on all frequencies of rotating machinery. Acceleration is usually used on high-speed rotating equipment. The signal is monitored by a transducer which creates an electrical signal. This electrical signal is transmitted to a meter that converts the signal to an output which can be read by the operator. This system can be used by a hand-held meter (Figure 4-13) held up to a remote unit that can receive signals from a transducer and be monitored in a central location.

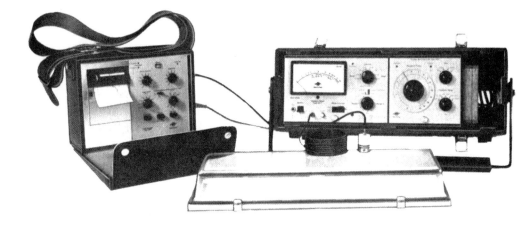

Some of the conditions that can be determined by vibration analysis are:

unbalance

defective bearings

misalignment

looseness

oil whip

bent shaft.

Any of these conditions are detrimental to the equipment life. Monitoring any increase in the acceptable levels of vibration will enable repairs to be effected before breakdown occurs.

Oil Analysis

This method of determining equipment wear is relatively new. It's a method of analyzing oil samples from lubrication oil or hydraulic oil to determine the amount of wear in a unit. The oil sample is drawn from a system and is then taken to an analyzer. The oil is examined to see what size and type of particles are in the oil. The more contamination there is in the fluid, the faster the wear will occur in the system. Seventy percent of all hydraulic system failures are attributed to contaminated oil. In lubrication systems, the effect of contamination has always led to rapid wear of the mechanical components.

The system used to analyze the particles usually consists of a particle counter and sizer (Figure 4-14). This equipment gives the average amount and size of the particles in the fluid and helps the technician decide whether the fluid is acceptable or needs filtration or perhaps even changing. The second part of the test is a spectro-

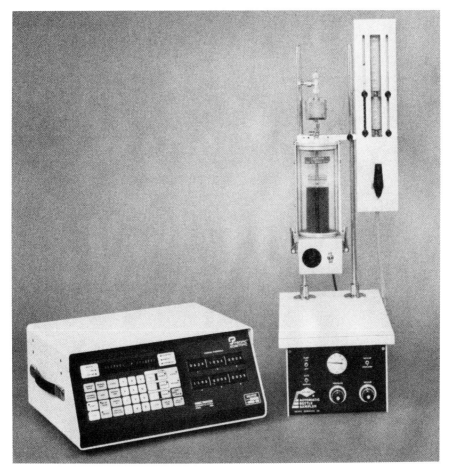

Figure 4-14. Particle counter and sizer. (Courtesy of HIAC/ROYCO.)

chemical analysis (Figure 4-15). The particles are examined to see what material they are made of. This will determine if the particles are dirt, wear material, or additives that have solidified. If the particles are just dirt, then filtration or changing of the oil is all that is required. If the particles are from the additives, then again changing the oil or adding additives is all that is required. If the particles are wear materials, then the system prints need to be checked to see which components the particles came from. If the amount of particles indicated excessive wear, then the component will be removed or repaired before a breakdown occurs.

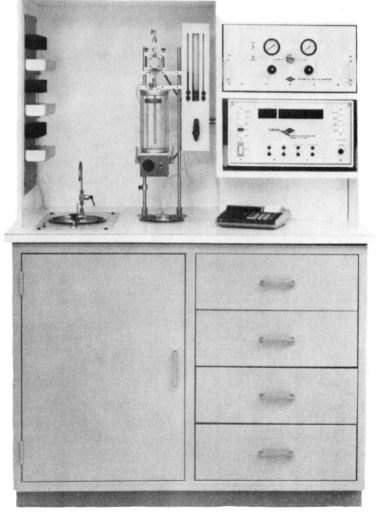

Figure 4-15.

This type of spectrochemical analysis is available from several firms, or with proper training and equipment can be accomplished in-house. It's one of the best inspection methods for operating hydraulic systems.

All of the systems discussed in this chapter have the potential to help pinpoint any preventive maintenance problem area. In fact, without the use of some of the systems discussed, a preventive maintenance program won't be much of a success. As technology continues to improve, so will the sophistication of the equipment used to monitor the condition of the operating equipment. This will have a positive impact on a preventive maintenance program.

An additional use for nondestructive testing is in the area of quality control. The testing is used to ensure that all products being manufactured are of good quality. This does not mean that each component must be inspected, but that a random sampling of the products be conducted. If several are found not to meet the minimum standards, then the entire group becomes suspect. When this occurs, individual inspections will be necessary.

5 Computerized Maintenance

The Importance of Record Keeping

Any maintenance organization is only as good as its record keeping. If clear, accurate records are not kept, how can the organization be managed? How can cost be calculated? How can work be charged to the right account? When was the last time the equipment broke down? What are the most used parts, ones that should be stocked in the store room? How much overtime was worked last month? Does manpower need to be increased or decreased?

Role of the Computer in a PM Program

With poor record keeping, some of those questions would be impossible to answer accurately. Even with good record keeping, it may take several hours to several days to derive an accurate answer. This is where the computer has an important application. The computer can be used to monitor any phase of the maintenance organization. Some of the records or duties the computer can chart are as follows.

1. Weekly work schedule of employees.
2. Creation of any needed work order (PM or corrective maintenance).
3. Scheduling of all PM inspections and repair.
4. Report on status of any work in progress.
5. Instant retrieval of all backlog work orders.
6. Tracking of all carry-over orders (from turn-to-turn or day-to-day).
7. Equipment history records.
8. Average time between failure and average downtime for the equipment.
9. Instant recall of any failure analysis reports.
10. Ratio of preventive maintenance to corrective maintenance.
11. Employee performance evaluations.
12. Equipment inventory.
13. Master equipment records.

As the list of possible computer usages is considered, the thought comes to mind: can't the same thing be accomplished by use of the card system or other form of record keeping? And the answer to that question would have to be: yes, it could. Then where is the advantage? This comes in time and number of employees required to keep records. It would take considerable time and man hours to perform all of the above-listed functions with a standard card system. However, with the computer, one individual trained to input the material can keep all of the records up to date from the information obtained from the work orders and maintenance reports received from the workforce and supervisors. This eliminates the problem of everyone wanting to keep records their way and no one really having any accurate idea of how the records are organized.

Cost is the next consideration of the computerized system. This is where the system is very flexible. If the installation is for a small maintenance organization, one microcomputer would be sufficient (Figure 5-1). The system should be large enough to handle the work load yet small enough to make the initial cost reasonable. The computer is accompanied by a system for data storage — usually disc drives (Figure 5-2) — and a method for receiving a printed copy of all material — a line printer (Figure 5-3). This basic equipment is suf-

Figure 5-1. CRT and keyboard of a microcomputer.

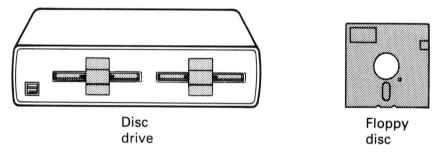

Disc
drive

Floppy
disc

Figure 5-2. Disc drives.

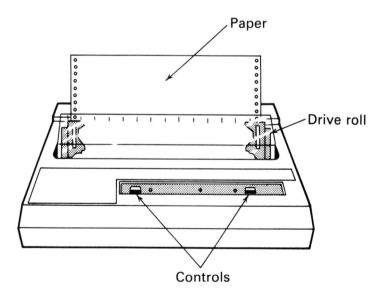

Paper

Drive roll

Controls

Figure 5-3. Line printer.

ficient to handle a small organization. As the size of the organization increases, so will the size of the computer system. But as the cost increases, so also will the savings from using the system.

In a large maintenance organization, it's possible to have a system as depicted in Figure 5-4. The computer system would be set up as follows. The central computer is in the main office. Here the maintenance planners, maintenance superintendent, and the maintenance engineers can have access to all records from all departments. The individual departments would have a computer relay terminal that is fed from the main computer. All inputs on the terminal would be available at the central location. This makes all the maintenance activities from all departments available to be charted or read at the supervisors' or engineers' convenience. The chance of error in written records or those transmitted over the phone is dramatically reduced.

The advantage to the field supervisor is that the computer is also tied into the central stores to allow almost instant access to spare parts information (Figure 5-5). The computer should also be tied into engineering (Figure 5-6) to allow a copy of any engineering

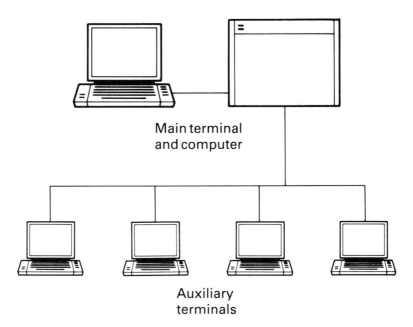

Main terminal
and computer

Auxiliary
terminals

Figure 5-4. Main terminal and satellite terminals.

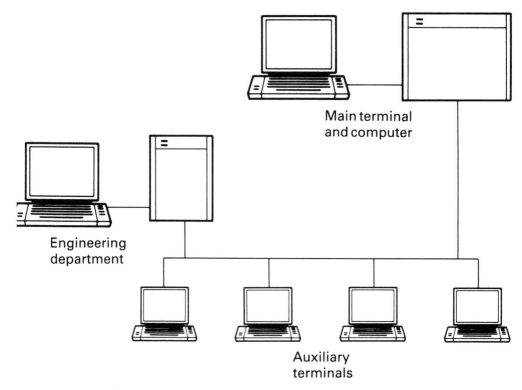

Figure 5-5. Maintenance organization tied to engineering.

blueprint to be consulted on the computer screen, or a copy to be printed from the engineering file. This eliminates the problem of keeping all prints up to date since they're all consulted from a central location and only one change will have to be made. Then each department wouldn't have to be notified, because the next time they looked at the print or copied it, the change would already be in place.

Computerized Troubleshooting

One of the most recent developments is computerized troubleshooting — at this time, the ultimate in computerized maintenance. It's also referred to as **fault diagnosis**. This method utilizes the plant monitoring equipment to the maximum (Figure 5-7). All inputs from a piece of equipment are tied into a central computer. In-

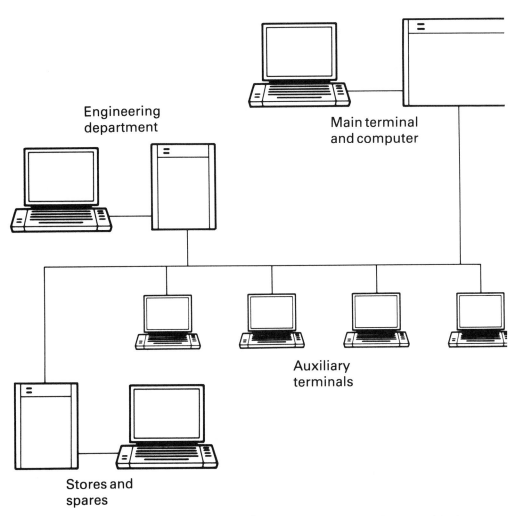

Engineering
department

Main terminal
and computer

Auxiliary
terminals

Stores and
spares

Figure 5-6. Maintenance organization tied to engineering and stores added.

cluded are operation and control sequences. When a fault occurs, the repairman goes to the equipment and plugs a hand-held monitor into a terminal on the equipment. By a series of yes/no questions, the computer leads the repairman to the problem, or he/she can troubleshoot from the central CRT.

These systems must be simple enough for a craftsman to understand and operate with a minimum of training.

In order for technicians to understand computerized systems, the

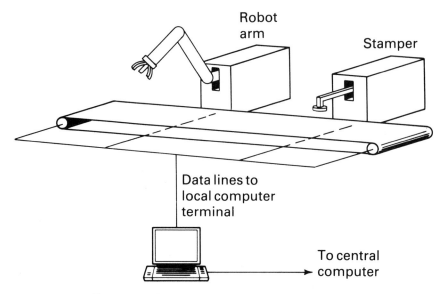

Figure 5-7. Plant monitoring and troubleshooting.

systems must be "user friendly," or easy to use. They must be flexible in operation so that they may be used in a variety of situations. This equipment doesn't require the upgrade of all employees to the level of advanced technician. Only very complex problems would warrant the use of technicians on shifts.

While this seems rather futuristic, computerized troubleshooting is becoming commonplace in industry. Think back ten to fifteen years. Who would have dreamed we would have the sophisticated robotic equipment we have today? And who is to say what technological advancements will be made in the next decade?

While no computer can think independently as can a human being, it can perform routine and repetitive tasks faster and with more precision. While the computer will never replace the maintenance supervisor in the making of logical decisions, or employees in troubleshooting and repair, it can be a useful tool to both groups. If it's used as a tool for getting the tasks accomplished, it can be a valuable asset to any maintenance organization.

II

Equipment Inspection, Maintenance, and Troubleshooting Guides

6 Mechanical Drives

Bearings

Bearings form the most important part of a mechanical drive system. Every shaft rotates or reciprocates in some form of a bearing.

Bearings are classified in two very broad categories: **plain (sleeve)** or **rolling element.** Although wear patterns differ between the two types, the three most common indications of a problem in a bearing are heat, noise, and vibration.

Overheating

What is heat? How hot is too hot? What is the normal operating temperature? The average bearing operates at 40 to 50°F above the ambient temperature surrounding the bearing. Any temperature reading above this level should be a cause for close inspection until the problem is located and corrected.

How are bearings checked for temperature? The obvious answer is a thermometer. However, sophisticated temperature monitoring devices that are currently marketed may provide an easier alternative than measuring each bearing temperature by hand. These devices may be as simple as a hand-held digital thermometer (Figure 6-1) that measures the bearing's temperature.

The hand-held thermometers are equipped with a probe that can be inserted into the grease hole in a bearing housing until it comes in actual contact with the outer race of the bearing. This gives a more precise reading than one that merely reads the outside housing temperature on the bearing.

Another alternative is a permanent transducer mounted on a bearing with a fitting for a hand-held meter to be connected to it. This enables the inspector to obtain an accurate reading with a minimum amount of time. If manpower is a problem, it's possible to run a feed from the transducer to a central control room, where one person could read the temperature of many bearings in a short period of time. The most sophisticated monitor would set off an alarm from a central location when the temperature reached a predetermined level (Figure 6-2).

Finding the Cause of Overheating: Determining that the bearing is hot is only part of the inspector's problem. The second part is determining what is causing it to overheat. No matter which of the following factors causes heat buildup, it must be corrected. If it is al-

Figure 6-1. Hand-held digital thermometer. (Courtesy of The Pyrometer Instrument Co., Inc.)

Figure 6-2. Central monitor system. (Courtesy of Brüel and Kjaer Precision Instruments.)

lowed to progress, rapid wear and eventual failure will occur. Let's look at the various causes of excessive heat in a bearing.

Lubrication: If the lubricant becomes suspect during inspection, it is usually because of one of three reasons:

1. wrong type of lubricant
2. insufficient lubricant
3. excessive lubricant.

The **wrong type of lubricant** is one of the most difficult problems to determine. All the lubricant levels and seals may appear to be in good condition. The only way to determine if it's the wrong type of lubricant is to keep accurate records of what's being used. When a change of type or even manufacturer is made and problems begin to develop, it'll be easier to determine the problem. If no record is kept, then the change may never be noticed unless someone has a good memory.

If the unit has just been installed and is starting up, it's best to consult with the manufacturer to consider their recommendations. They'll usually be able to determine the correct lubricant for the installation, taking into consideration all of the conditions surrounding the equipment in conjunction with the operating characteristics.

If the inspector determines that the lubricant is the right type, attention should next be given to the *lubricant level*. If the **level is too high,** the heat may be generated from churning the lubricant. Churning of the lubricant is actually high-speed stirring. This is occurring so fast that friction develops in the lubricant itself. The friction of course builds up heat which is transferred to the bearing and housing (Figure 6-3). The higher the temperature rises, the worse the situation becomes. It will finally reach a point where the lubricant fails and the bearing is destroyed.

Too little lubricant is just as big a problem as too much. The bearing, in theory, never has contact between the races and the rolling elements. The lubricant is supposed to build a film wedge to separate the races from the rolling element (Figure 6-4). When there

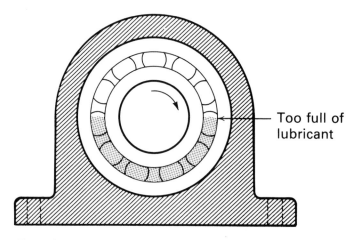

Too full of lubricant

Figure 6-3. Lubricant level too high in the bearing and housing.

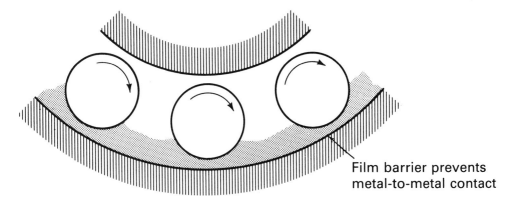

Film barrier prevents
metal-to-metal contact

Figure 6-4. A film of lubrication.

is insufficient lubricant, the film wedge cannot be formed and metal-to-metal contact occurs, and heat is generated. When the contact occurs, the heat generated is so intense that welding of the rolling elements occurs. If rotation is continued, the weld is broken loose and then occurs again. This process repeats itself until the bearing is destroyed.

Installation Practices: Poor installation practices are also a cause of heat in a bearing. The four common areas are: poor shaft fits, poor housing fits, misalignment, and rubbing seals.

Poor shaft fits occur when the shaft is either too large or too small. If the shaft is too large, the bearing's inner race is expanded to be installed on the shaft. A certain amount of expansion is necessary and is taken into consideration by the manufacturer when it's designed. Excessive expansion causes the internal geometries of the bearing to be distorted (Figure 6-5). This will remove the necessary internal clearances between the rolling elements and the raceways. When the bearing is rotated under load, the lubricant barrier cannot be formed and the welding and tearing process begins. This generates even more heat and the bearing self-destructs.

If the shaft is too small, the bearing's inner race is not expanded at all, and it has no holding power on the shaft. It can be compared to a rubber band, which must be stretched slightly if it's to have any holding power at all. So if the race is not expanded to its proper size, it slips on the shaft. As it continues to slip, wear occurs. As this

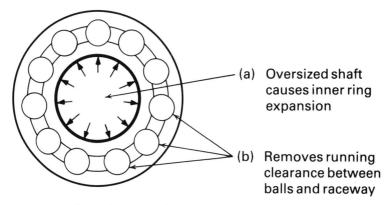

(a) Oversized shaft causes inner ring expansion

(b) Removes running clearance between balls and raceway

Figure 6-5. Excessive expansion in the bearing.

wear occurs, frictional heat is generated. The combined wear and heat rapidly destroy not only the bearing, but also the shaft (Figure 6-6).

Poor housing fits are essentially the opposite of shaft fits. If the housing fit is too loose, then it slips. The slippage causes wear and heat, and destroys the bearing and the inside of the housing. If the housing fit is too tight, the outer race is compressed (Figure 6-7). This condition reduces the internal clearances in the bearing. As the bearing is run, the film of lubrication won't develop, and the welding and tearing procedure begins, and the bearing is rapidly destroyed.

Proper shaft and housing fits are essential if the bearing is to run

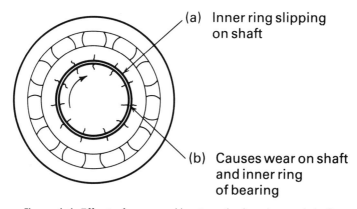

(a) Inner ring slipping on shaft

(b) Causes wear on shaft and inner ring of bearing

Figure 6-6. Effect of wear and heat on the bearing and shaft.

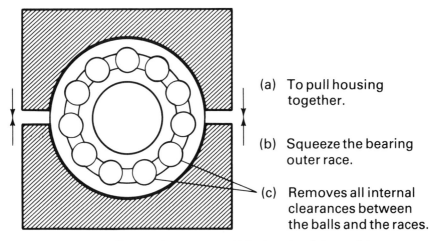

(a) To pull housing together.

(b) Squeeze the bearing outer race.

(c) Removes all internal clearances between the balls and the races.

Figure 6-7. Effect on the outer race if the housing fit is too tight.

problem-free. To obtain a copy of recommended fits, consult your local bearing supplier.

Misalignment is another installation problem that will cause the bearing to generate heat. If the shaft that the bearing is supporting isn't properly aligned, the bearing will wear in a pattern for which it was not designed. The rolling element is given a certain path to follow. If the bearing is misaligned, the rolling element will try to run in a new path. This change in the internal geometries (Figure 6-8) of the bearing won't allow proper lubrication or loading. This will result in metal-to-metal contact (under load) in the bearing, and in the

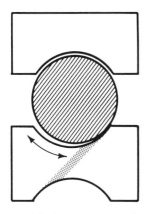

Ball path runs from one side to the other (instead of in the middle) when the bearing is misaligned.

Figure 6-8. If the bearing is misaligned, the path of the ball runs from one side to the other, instead of in the middle.

welding and tearing of the contacting surfaces. Again, rapid deterioration and failure result.

Rubbing seals (Figure 6-9) cause excessive heat buildup when they rotate with the outer race of the bearing and rub on a shaft shoulder or perhaps run at too high a speed. The friction of the seal on the interference causes heat buildup and eventually destroys the seal, and perhaps contaminates the bearing.

When an inspector finds a bearing that's overheating, arrangements should be made to carefully inspect it at a time when it isn't operating. If an obvious problem is not found, disassembly of the unit may be necessary. Once the cause has been determined and the bearing changed (if necessary), the unit should be reassembled and

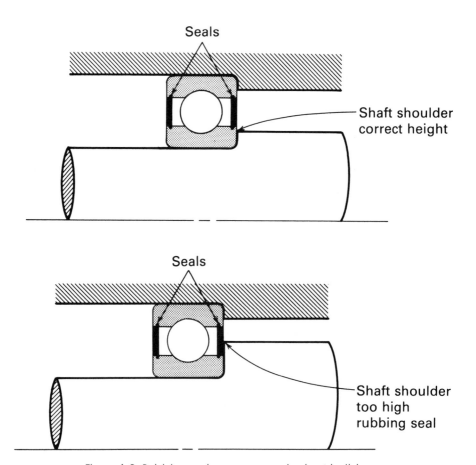

Figure 6-9. Rubbing seals cause excessive heat buildup.

carefully monitored until it has been determined that the problem was eliminated.

Noise

Noise in a bearing can indicate a developing problem. In most cases, when noise develops, it's accompanied by heat or vibration. But noise alone can call the inspector's attention to the bearing's problem without the inspector being too close. Some of the common problems that cause noise in a bearing are:

1. loose shaft or housing fit
2. tight shaft or housing fit
3. contamination
4. internal wear.

Loose and Tight Shaft and Housing Fit: Poor housing and shaft fits cause noise as well as heat (which was previously described). The noise is caused by the slipping action of a loose fit or the lack of internal clearances caused by a tight fit. It the fit is loose, there is a low rumbling sound, because the race is moving when it's under load. If the fit is too tight, it'll be accompanied by a high-pitched, squealing sound, because the bearing has insufficient internal clearances and the sound is the rubbing of metal on metal at high speed and load. These conditions are indications of trouble in the bearing, and the inspector should know that rapid wear will result unless prompt action is taken.

Contamination: Contamination in a bearing can cause excessive noise because the contamination interferes with the normal contact path of the rolling element (Figure 6-10). Contamination is actually a

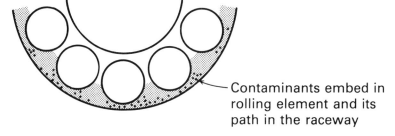

Contaminants embed in rolling element and its path in the raceway

Figure 6-10. Contaminants in rolling element path.

twofold problem. You're trying to keep the dirt out of a bearing and at the same time you're trying to keep some form of lubrication in the bearing. There are two basic ways for dirt to enter a bearing. It can enter when the bearing is handled during installation (Figure 6-11), or it can enter when the bearing is in operation (Figure 6-12) — which becomes a sealing problem.

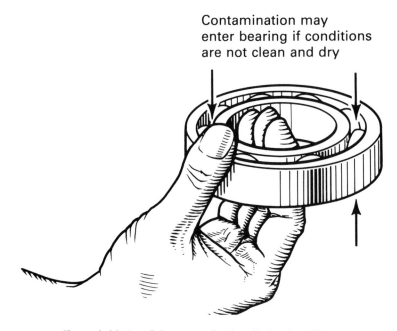

Contamination may enter bearing if conditions are not clean and dry

Figure 6-11. Possible contamination during handling.

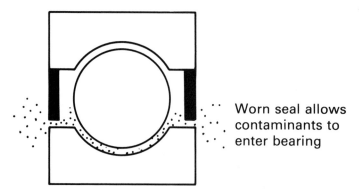

Worn seal allows contaminants to enter bearing

Figure 6-12. The seal can allow contaminants to enter the bearing.

Regardless of how the bearing has become contaminated, it will have the same signs. If the noise is caused by contaminants entering into the bearing, there will usually be markings in the raceways. These take the form of scratches or pits if the contaminants are very large. If the contaminants are very small, they'll act like a lapping compound and wear away the raceways and the rolling elements (Figure 6-13). This increases the internal clearances and allows the excessive noise. One additional point to inspect is the condition of the sealing device around the bearing. If it's leaking or worn away, chances are contaminants are entering into the bearing and causing the problem.

If the contaminant is some form of corrosive or water, the bearing may look like it has rusted. This is very harmful, especially if it forms in the raceway or on the rolling element (Figure 6-14). This will interfere with the lubricant barrier and drastically shorten the life of the bearing. If it forms on the inner race or the outer race, it could possibly interfere with the proper contact of the bearing with the housing or the shaft, allowing slippage to take place. Either condition will be evidenced by increased noise during operation.

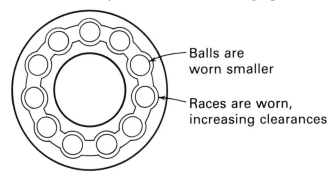

Balls are worn smaller

Races are worn, increasing clearances

Figure 6-13. Small contaminants wear away the raceways and the rolling elements.

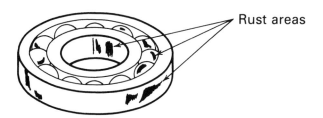

Rust areas

Figure 6-14. Rust in the raceway and rolling elements.

Internal Wear: Internal wear in a bearing may also be a cause of noise during operation. The wear inside the bearing can be evidence that the bearing is ending its useful life and is failing due to fatigue; but since only about 5% of all bearings fail for this reason, it's likely to be for another reason. The most common reason is that the bearing has become contaminated and the contaminants have accelerated the internal wear. In either case, it will be necessary to change the bearing.

Vibration

Vibration in a bearing can have several causes. Some foreign matter entering the bearing will cause it to vibrate as the wearing process occurs in the bearing.

Any incorrect shaft or housing fit that changes the internal geometries within the bearing will cause it to vibrate. Again, the cure will be to check the shaft and housing fits.

One cause of vibration that has not been previously considered is the matter of balance. If a shaft or some part of a drive is out of balance, it can cause the bearing to vibrate excessively. It will probably require the use of a vibration meter to locate the problem. The solution depends on the problem and its location in reference to the bearing.

While there may be some specialized bearing problems, most will fall into the three main categories of heat, noise, and vibration. It's very possible, in fact likely, that the symptoms will occur in combination, with possibly all three being present. It will take experience and knowledge on the part of the inspector to determine the true cause of the problem.

Chain Drives

We'll discuss here roller chains, but the same principles have application in other types of chains as well, including silent chains and conveyor chains. In all chains there is some form of mechanical joint (Figure 6-15) where a bending action takes place. It's at this bending point that most wear occurs, and thus this should be one of the main inspection points. In addition, examining other wear patterns on the chains will indicate problems in the drive.

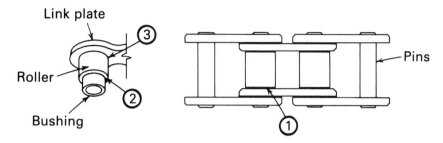

Figure 6-15. Joints in chain drives.

Chain inspections should fall into four main categories:

1. normal wear
2. tension
3. sprockets
4. lubrication.

Normal Wear in Chain Drives

Normal wear is the process of wear that takes place in the chain joint from continual flexing. Standard roller chain pins and bushings are case hardened. This means only the outer parts are hardened, and the inner parts are of correspondingly softer metal (Figure 6-16). You could compare them to an M&M® candy — coated on the out-

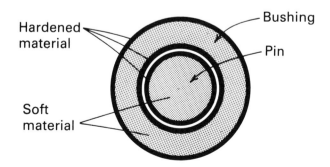

Figure 6-16. Standard roller chain pin and bushing.

side and softer inside. Once the outer shell is worn off the chain part, the softer internal parts will wear much more rapidly and can actually fail very shortly. This is an important point for an inspector to examine. As this wear occurs, one of the most obvious signs is that the chain appears to be stretching or elongating.

Elongation occurs because the bushings and pins are wearing, and thus as the material is removed it increases the amount of play in the chain joint. This increased play will allow the chain to increase its length. How can the elongation be checked? Any time the elongation is more than 3% of total chain length, it's worn out and can fail at any time thereafter. So if a chain had a length of 100 inches when installed, and on inspection it's stretched out and is 103 inches in length, it's worn out (Figure 6-17). The outer hardened shell of material will be worn away, leaving the softer material exposed. This material will wear at a much faster rate, weakening the load-carrying capacity of the chain. The inspector should recommend changing the chain at the 3% elongation point to prevent a breakdown.

One other practice in industry that accelerates wear in a chain

1. Find a section of chain under tension.
2. Measure length of a given number of pitches.
3. Compute normal length.
4. If length is 3% oversize, change the chain.

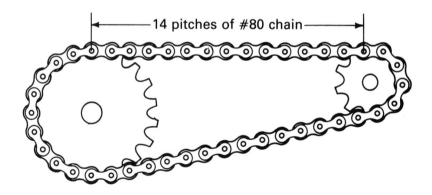

Example
1. 14 pitches of #80 chain = 14 inches
2. If it measures 14.4 inches, it needs to be changed.

Figure 6-17. Determining elongation in a chain drive.

drive is replacing links in a chain. If the chain does break, many times the repair will consist of removing the damaged link and replacing it with a new link. If the chain was worn near the 3% elongation point, this practice may destroy the drive. If the point is considered that the chain is worn near the 3% level, the pitch of each link is longer than when new. If a link with a different pitch is installed, the chain will engage the sprocket at a different time (Figure 6-18), and will create a shock in the chain as it enters or leaves the sprocket. This shock will cause more wear to occur on the links on either side of the repaired area, which will result in failure in the area of the new link. *Putting in more links will only increase the problem.* It is best to recommend changing the chain.

Tension in a Chain Drive

Tension problems in a chain drive can be divided into two classes: too much or too little. Tension in a chain drive should be measured on the slack side of the drive. A straightedge should be run from one sprocket to another (longer center distances may require the use of a wire or string). At the center of the drive, a 2–3% deflection should be present in the chain (Figure 6-19).

If there is less than that amount, the chain is too tight and should be loosened. If it isn't loosened, then as the chain enters the sprock-

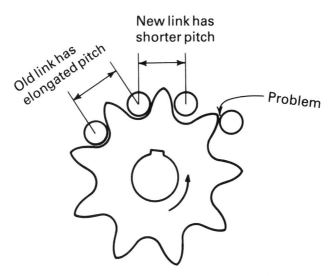

Figure 6-18. New link in old chain causes chain to engage sprocket improperly.

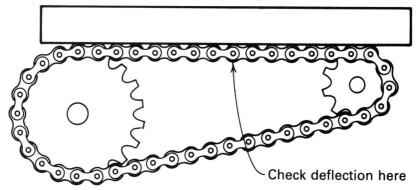

Figure 6-19. Checking tension in a chain drive.

et it will pull on the upper part of the tooth. This is not where it's designed to engage the sprocket (Figure 6-20). If this condition continues, it will wear on the upper portion of the tooth, causing excessive loading, possibly breaking the tooth off the sprocket. If this doesn't happen, then the increased loading will wear the chain at a rate much higher than normal, resulting in a dramatically shortened life.

Too little tension can be just as damaging. When the chain enters the sprocket with too little tension, it'll engage the tooth late, thus riding high and being pulled into the tooth space. As wear occurs on both the sprocket and chain, it'll ride even higher on the sprocket

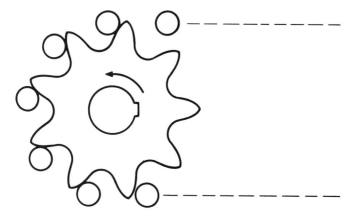

Figure 6-20. Chain riding on top of sprocket teeth because chain is too tight.

tooth. As this occurs, the chain will finally begin slipping a tooth as the chain engages the sprocket (Figure 6-21). This creates a shock in the drive. As this continues, the drive literally beats itself to death. If this condition is observed, repairs should be effected as quickly as possible. There is no set number of times the chain will absorb this type of shock. It can actually break at any time. Tension should be an important point in inspection of any roller chain drive.

Sprockets

Sprocket inspections are important to determine when sprockets should be replaced. If a chain is run on a worn sprocket, accelerated

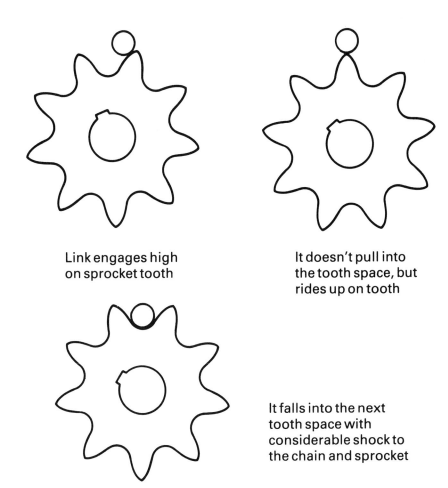

Link engages high
on sprocket tooth

It doesn't pull into
the tooth space, but
rides up on tooth

It falls into the next
tooth space with
considerable shock to
the chain and sprocket

Figure 6-21. Effect of too little tension on the chain.

wear occurs. But this doesn't mean each time a chain is replaced that the sprocket should be replaced also. A sprocket kept in good condition can outlast three or more replacement chains. The problem occurs when the sprocket is worn. What do you look for?

The shape of the tooth is an important consideration. The tooth has a small, circular-shaped area which the chain is to ride in when in contact with the sprocket. If it doesn't ride in this area, it will wear the sprocket in another area causing it to become hook shaped (Figure 6-22). Once this occurs, it's best to change the sprocket, for it will wear the chain faster than normal. The hook-shaped tooth is also caused by normal wear. But with normal wear, the hooked area will usually be lower on the tooth. It still indicates a need for the sprocket to be changed (Figure 6-23).

Another location to look for wear on a chain sprocket is on the sides of the teeth. Wear on the side of the teeth indicates an alignment problem between the sprockets. If the sprockets are misaligned, a chain will rub the teeth on one side of one sprocket and on the opposite side of the other sprocket (Figure 6-24). Another inspection point for misalignment is on the inside of the link plates of the chain. A corresponding wear occurs in this location as well. Alignment problems may also be caused by defective bearings allowing the sprockets to turn or move, simulating the conditions of misalignment (Figure 6-25). It may be best to observe the drive while it's in operation to be sure the problem was diagnosed correctly.

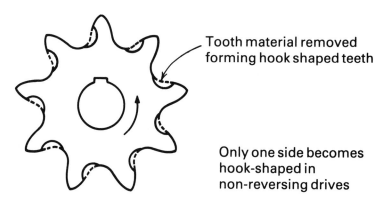

Tooth material removed forming hook shaped teeth

Only one side becomes hook-shaped in non-reversing drives

Figure 6-22. Effect of a worn sprocket on the teeth. Only one side will become hook shaped in nonreversing drives.

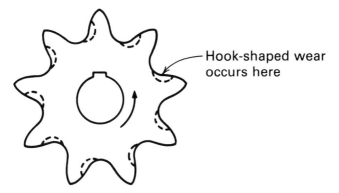

Hook-shaped wear occurs here

Figure 6-23. Normal sprocket wear.

If inspected properly, the sprocket can give good indications of problems. An alert inspector will be wise in including the sprockets in any chain inspection.

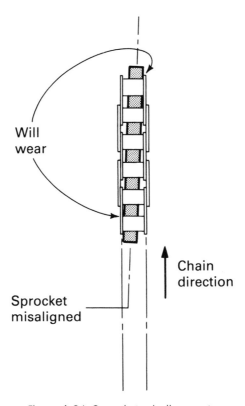

Will wear

Chain direction

Sprocket misaligned

Figure 6-24. Sprocket misalignment.

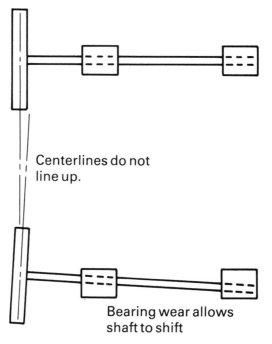

Figure 6-25. Bearing causing sprocket misalignment.

Lubrication

Lubrication is the single most important factor in any chain drive. It should be carefully examined during the inspection. A chain that isn't lubricated wears three-hundred times faster than one that's well lubricated — a good reason for everyone to pay attention to chain lubrication. But just because the chain has some oil or grease on it doesn't mean that it's lubricated.

As discussed earlier, the main wear in a chain takes place in the pin and bushing joint area of a chain. It's imperative that the lubricant penetrate this area if it's to be effective (Figure 6-26). If the lubricant cannot get into the area and provide a metal-separating barrier, wear will occur. Thus, the use of heavy, thick oils or greases will not fulfill the lubrication needs of a roller chain.

How can you tell if the lubricant is effective in penetration of the joint? Most roller chains have a connecting link or an offset link for ease of installation and removal. To inspect the lubricant's effectiveness, remove this link. Inspect the pins. If they have a smooth, polished look, the lubricant is penetrating the joints and preventing

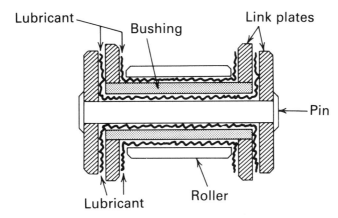

Figure 6-26. Roller, pin, and bushing lubrication. Oil drips on inside of chain.

wear. If they're dark brown or coated with red deposits and have a scored appearance, the lubricant isn't getting the job done and should be changed.

If it's the correct lubricant, then perhaps the method of application is wrong. There are only four main types of lubrication systems: drip, bath, slinger, and spray.

The drip and spray methods have the same problems — direction of oil flow. It's important, especially on higher speed drives, that the oil is directed to the inside of the chain strand (Figure 6-27); then centrifugal force forces the oil into the chain joint as it goes around the sprocket. If it's directed to the outside of the chain, then centrifugal force will throw the oil off the chain before it can lubricate. The direction of the oil sprays should be a primary inspection point.

Disc and bath lubrication systems have a common problem also — the oil level. If the oil level is too low, the chain won't receive adequate lubrication (Figure 6-28). If it's too high, the chain may churn the oil, overheating it. While the lubricant may be penetrating, the oil's lubricating ability is destroyed by the churning and the resulting overheating. Both conditions should be carefully checked by the inspector.

Roller chain drives are a very dependable and economical form of a positive drive. Careful and timely inspections will enable the drive to last an indefinite period of time. Ignoring the inspection points in a drive will reduce its life, making the use of it very costly.

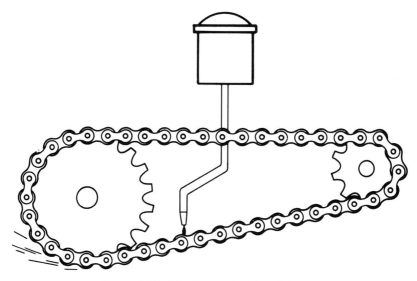

Figure 6-27. Oil drips on inside of chain.

Belt Drives

It's easy to make an inspection of belt drives. Most all signs of trouble are easily spotted and diagnosed. The problems in belt drives can be divided into the following categories:

- tension
- environmental
- installation damage
- overloads.

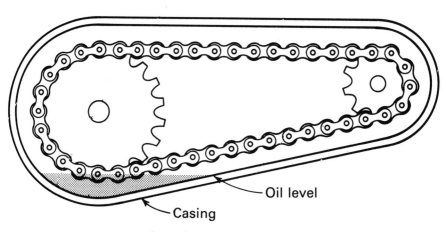

Figure 6-28. Correct oil level.

Tension Problems

Tension problems are divided into two classes: too much and too little. Too much tension in a belt drive will leave the belts feeling rigid, unyielding. The bearings supporting the shafts will be more heavily loaded and will usually run hotter than normal (Figure 6-29). If this is the case, the belts should be retensioned. The best method is to use a tensioning tool, which is available from the belt distributor (Figure 6-30).

The tool will have instructions on installation and tensioning of all three main types of V-belts. This is important because the wedge-type belt needs more tension than the other two types. If the belt is too loose slippage will occur. This may or may not be accompanied by belt squeal. V-belts can usually slip up to 20% before the squealing takes place. Slippage is important to stop because it accelerates wear between the belt and the sheave, cutting the sidewall on the belt and dishing out the side of the pulley sheave (Figure 6-31). A good, quick method of checking for slippage is to remove the guard and put a chalk mark on the belt, particularly in a multiple belt drive (Figure 6-32). As the belts rotate, any change of position is easy to spot.

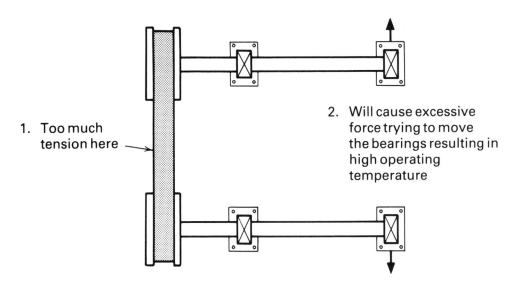

1. Too much tension here →

2. Will cause excessive force trying to move the bearings resulting in high operating temperature

Figure 6-29. Excessive belt tension.

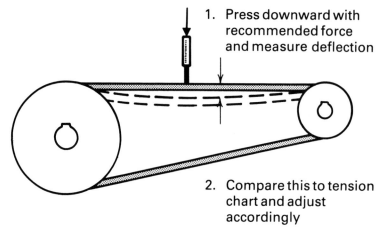

1. Press downward with recommended force and measure deflection

2. Compare this to tension chart and adjust accordingly

Figure 6-30. Proper use of tension tool.

Environmental Conditions

The next inspection area should be the environmental conditions. Any **foreign matter** that gets into a belt drive accelerates wear on a belt. If dust or dirt gets in between the belt and the sheave sidewall, it acts as an abrasive cutting agent, wearing away at the side of the belt (Figure 6-33). This will continue until the belt is worn sufficiently to break under load. If inspection reveals a rapidly wearing sidewall, and no slippage is present, this would be a good point to check. Dust and dirt can also wear on the pulley sidewall, dishing it out. This occurs along with belt wear.

If the belt is merely replaced, the worn sheave will rapidly wear the new belt. The sheave should be inspected for wear. This inspec-

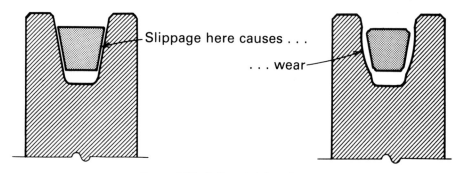

Slippage here causes . . .

. . . wear

Figure 6-31. Belt and sidewall wear.

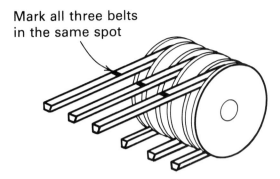

Mark all three belts
in the same spot

Figure 6-32. Chalk marking multiple belts to check for slippage.

tion is best performed using a sheave gauge. It lets the inspector know if the wear exceeds the allowable limit. If the wear does exceed the allowable limit, it should be noted so the sheave pulley can be changed as well as the belt.

Oil and heat can affect a belt's life also. Most petroleum-based oils cause a chemical reaction with the rubber in the belt, causing it to swell and become very sticky (Figure 6-34). This condition will progress and break down the load-carrying members in the belt, causing premature failure. Any time oil is found on the belts during an inspection, it should be noted and a way found to prevent the problem from recurring.

Excessive temperatures in a belt drive result in hardening of the belt, and cracking will occur in the belt (Figure 6-35). The reason for

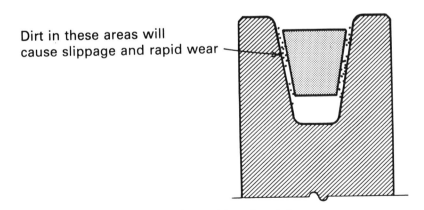

Dirt in these areas will
cause slippage and rapid wear

Figure 6-33. Dirt on sidewall and sheave.

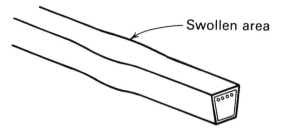

Figure 6-34. Swollen belt.

this is that during manufacture the belts are cured at a given temperature condition for a specified time period. If the belts are in use in a temperature above 130°–140°F, the curing process begins again. When the belts become overcured, they will crack when flexed around the pulley. There are special belts designed for this temperature range if the temperature cannot be lowered. They are more expensive, but they may have to be used if the temperature cannot be lowered. The important thing is to keep the belts properly protected. Keep all foreign matter out of the drive and keep all fluid off the belts. If guards must be redesigned, then have this job assigned. If properly protected, the belts will deliver a considerably longer service life.

Installation Damage

Damage during installation is one of the most overlooked areas in belt inspections. A cut in the outer cover of the belt can be caused by someone prying the belt on during installation (Figure 6-36). A cut can also be caused by an object falling into the drive, but this is rare if the belt is properly guarded.

If the belt has a tendency to turn over in the drive, it's usually caused by having broken internal cords. This can be caused by not

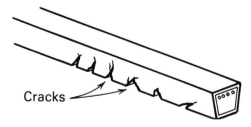

Figure 6-35. Belt cracking.

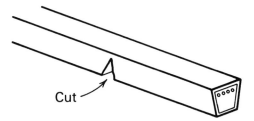

Figure 6-36. Results from prying belt on during installation.

reducing the center-to-center distances of the pulleys to install the belts and prying them on instead.

Alignment of the belt sheave is very important. While a belt is more flexible than a chain, it will still wear faster if the sheaves are not aligned. The belt will be pulled against the edge of the pulley sheave, causing wear on the sides of the belt as well as wear on the edges of the pulleys (Figure 6-37). The best way to make a field check for alignment is to use a piece of string and run it along both pulleys to ensure that it touches on the edges of both flanges (Figure 6-38). If it doesn't, proper action should be taken before damage occurs.

Overloads

Overloads are the last inspection category. A prolonged or shock overload will cause breakage of the belts, because overloads usually

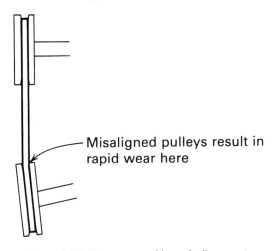

Figure 6-37. Wear caused by misalignment.

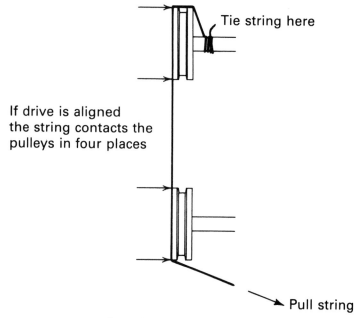

Figure 6-38. String alignment.

cause slippage of the belt, accelerating the normal wear. This slippage can be checked by the chalk method if overloads are suspected. If the drive has been properly tensioned and slippage still occurs, then the belt should be checked to ensure that it's the proper belt for that particular application. If the drive is overloaded, then changing the pulleys or upgrading the types of belts used may be in order.

Following these brief inspection hints will enable the inspector to make a good analysis of any V-belt drive.

Timing Belts or Positive Drive Belts

Timing belts or positive drive belts are the second most common type of belt being used (Figure 6-39). These belts have teeth and run against a toothed pulley, allowing them to have the characteristics of a belt drive and the positive drive feature of a chain drive. Timing belt inspections should be concerned with tension and alignment.

Tension problems occur because there is either too much or too little tension. Too much tension causes excessive bearing wear and

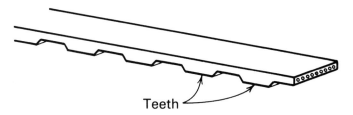

Teeth

Figure 6-39. Timing belt.

may also cause heat to be generated between the belt and pulley. Too little tension may cause the belt to jump teeth in the pulley (Figure 6-40). This causes shock loads in the belt and will eventually break the load-carrying members in the belt. This type of loading can possibly shear the teeth off the belt.

Misalignment will cause the belt to run to one side of the pulley or the other. Since the belt drive has at least one flanged pulley, the belt will show wear on the side where it's rubbing the flange on the pulley (Figure 6-41). It's possible for the belt to run up on the flange if the misalignment is severe enough. This may result in a cutting of the belt around the circumference (Figure 6-42). If this happens, it's best to change the belt and realign the pulley.

Most positive drive belts are not susceptible to environmental conditions, and can run in wet or oily environments.

If good inspections are made and problems corrected before the belts are damaged, such belts will provide an economical method of transmitting power.

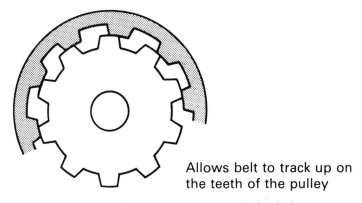

Allows belt to track up on the teeth of the pulley

Figure 6-40. Too little tension on timing belt.

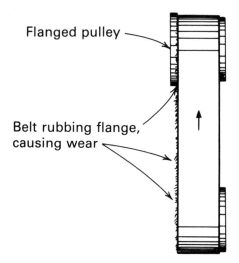

Flanged pulley

Belt rubbing flange,
causing wear

Figure 6-41. Flange wear.

Gears

Gear inspections are usually more difficult to carry out than inspections of the other drives, for they are usually enclosed in a housing or a casing. However, what we say about housed gears also applies to open gearing. Inspection of gears usually involves an

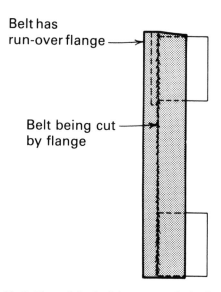

Belt has
run-over flange

Belt being cut
by flange

Figure 6-42. Cutting of the belt because of misalignment.

analysis of the gear tooth surfaces (Figure 6-43), because they tell the condition of the drive, along with pinpointing any problems that are occurring in the drive. Tooth surface problems may be divided into four classes:

1. normal wear patterns
2. lubrication problems
3. alignment problems
4. shock or overloading problems.

Normal Wear on a Gear

Normal gear patterns on gear teeth are caused by proper wearing-in of the gear teeth. These patterns are evidence of slow removal of material from the tooth surface and should not shorten the expected life of the gear. This process usually removes the high asperities from the tooth surface and develops a smooth, fine surface (Figure 6-44). If a gear tooth is heavily loaded, the material may be removed more quickly from the tooth surface. This will shorten the tooth's life. The material may be removed in the form of flakes or perhaps pits (Figure 6-45). One solution is to use lubricant with a higher viscosity. The added film thickness may help to reduce loading on the tooth.

Another form of wear that may be present in the startup of a gear drive is **pitting**. Pitting occurs when the surface material of a

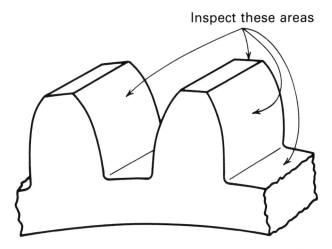

Figure 6-43. Gear tooth surfaces should be inspected.

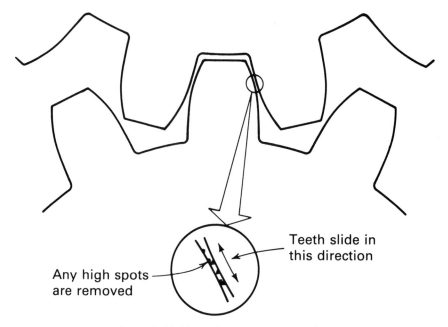

Any high spots are removed

Teeth slide in this direction

Figure 6-44. Normal wear on gear teeth.

gear is loaded beyond its endurance limit. Small parts of the surface that are being stressed break out and fall off (Figure 6-46). The process usually stops once the overstressed areas have been removed.

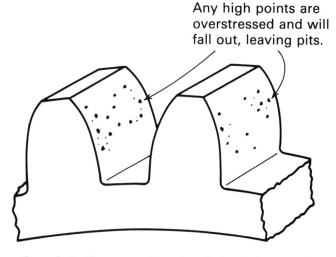

Any high points are overstressed and will fall out, leaving pits.

Figure 6-45. Wear caused by a heavily loaded gear tooth.

Wear occurs in this area;
asperities fatigue out,
leaving pits.

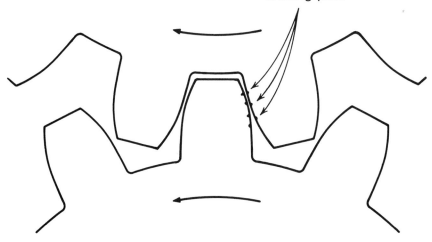

Figure 6-46. Pitting caused when the surface material of the gear tooth is loaded beyond endurance.

The tooth usually polishes up after a period of operation, and the life won't be drastically shortened.

If the pitting fails to stop, it indicates that the tooth may be overloaded. The smaller pits combine and larger pits are formed (Figure 6-47). This process continues until the tooth weakens enough to break off. The only solution is to harden the surface or reduce the loading.

Another form of pitting is called **spauling**. The pits are not deep and may be larger in diameter. This condition is also caused by heavy loading on the tooth surface.

Lubrication Problems

Lubrication problems in a gear drive cause a variety of conditions. If the lubricant is contaminated with foreign particles, the teeth will develop scratches. These scratches mark the teeth in the direction of the sliding action of the teeth (Figure 6-48). The larger the particles, the deeper the scratches. If found soon enough, the damage won't be severe enough to shorten the life of the drive. The best remedy is to filter the oil. If the oil is not sprayed but is in a splash system, it'll be necessary to change the oil in the gearcase.

Larger pits are formed by
combination of smaller ones.
This begins at the pitch line.

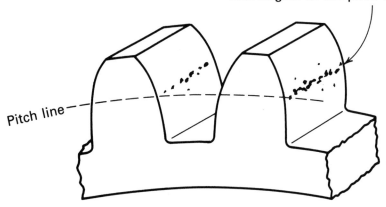

Pitch line

Figure 6-47. Larger pits are formed by combination of smaller ones. This begins at the pitch line.

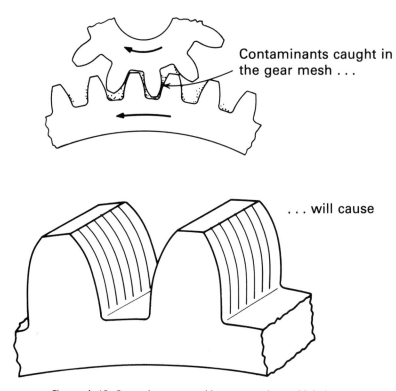

Contaminants caught in
the gear mesh . . .

. . . will cause

Figure 6-48. Scratches caused by contaminated lubricant.

A second lubrication failure is **scoring** (Figure 6-49). Scoring is a condition in which there is a rupture in the oil film barrier while the teeth are in mesh. This allows the surfaces to come in direct contact under load, and the parts of the two surfaces to weld together. As the teeth come out of mesh, the welds are broken, taking material from one tooth and leaving it on another. As the tooth goes back into mesh, the process is repeated until failure of the tooth occurs. The only solution is to use a lubricant that contains an extreme-pressure additive that won't allow metal-to-metal contact in the mesh area.

Another form of wear caused by lubrication problems is **corrosive wear**. Some corrosive element affects the tooth surface. This can occur because an outside material, such as acid, water, or chemicals, is in the gearcase. Not only outside material but also the lubricant itself can contain additives that become corrosive over a period of time. Constant watch must be kept on any gears that begin to develop this type of wear. A change of lubricant, with fresh lubricant being added, is the only cure for corrosive wear.

Rippling is another wear pattern that can be controlled by lubrication (Figure 6-50). Rippling is a wear pattern indicating that the

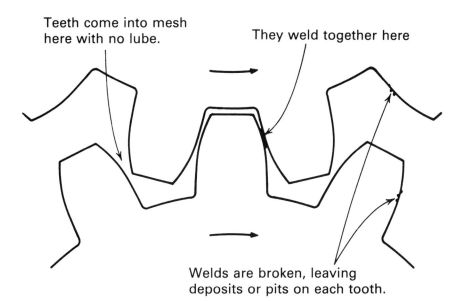

Teeth come into mesh here with no lube.

They weld together here

Welds are broken, leaving deposits or pits on each tooth.

Figure 6-49. Scoring caused by a rupture in the oil film barrier.

subsurface material in the gear is being moved. This explains the wave-type form that the tooth takes. Hardening of the tooth material or using a lubricant with an extreme-pressure lubricant can prevent this problem.

Alignment Problems

Alignment in a gear drive is critical if high contact stress is to be avoided. If the gears aren't correctly aligned in the vertical position, there will be contact between the tip of one tooth and the root of another (Figure 6-51). This results in rapid wear and failure. Correct spacing should be checked in all gear drives during any overhaul. If the alignment is off side to side, one side of the tooth will carry all of the load, resulting in rupturing of the oil film barrier (Figure 6-52). This will allow scoring of the surface to take place. If the misalignment is slight, the problem will correct itself. If there is considerable misalignment, it's possible that the tooth will become scored so much that it'll break off. If the misalignment is great, it may also overload the teeth beyond their stress levels so that they break off as they pass through the load zone. Alignment should be a key inspection point on any drive that has recently been disassembled.

Shock or Overloading

Shocks or overloads in a gear drive also leave signs on the gear teeth. Shocks or overloads are often great enough to cause **rolling and peening** of the tooth surfaces, a condition in which tooth material is moved in the direction of the sliding. This leaves fin-like pro-

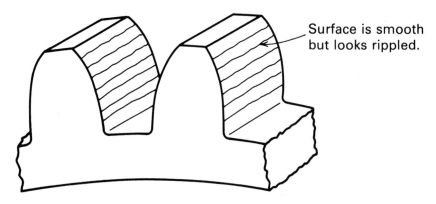

Surface is smooth but looks rippled.

Figure 6-50. Rippling.

1. Insufficient vertical
   clearance between
   gears . . .

2. . . . will cause excessive
   contact and loading
   here

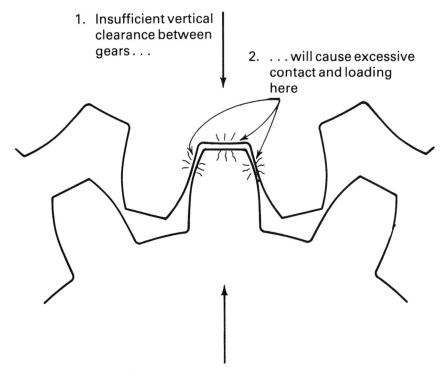

Figure 6-51. Problem caused by vertical misalignment of gear drive.

jections on the tooth surface (Figure 6-53). The tooth material is continually moved as it travels in and out of the load zone. The tooth profile is deformed and breakage will eventually follow. Some possible solutions include hardening the tooth material or reducing the loads.

Overloads will also cause the formation of **fatigue cracks**, caused by the tooth being heavily stressed (Figure 6-54). The flexing of the tooth will result in fatiguing of the tooth. A crack will form and progress as the flexing continues (a crack may also progress from a defective grinding crack or from a crack when the gear was quenched). As the crack progresses, it results in the eventual breakage of the tooth.

A severe shock load may break the tooth off in one overload. The difference between the two forms of tooth breakage will be hard to spot, but the overload will not leave the progressive crack as an indication of what happened.

Gear inspections are important if the cost of the drive is to be-

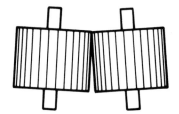

This type of misalignment . . .

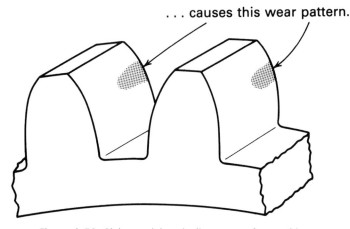

Figure 6-52. Side-to-side misalignment of gear drive.

come competitive. Gear drives cost more than the other drives, but they can last a lot longer than the others if they're maintained properly.

Couplings

Couplings, no matter what type, all have one thing in common: they need proper alignment. Any coupling that isn't aligned won't perform properly. No matter how flexible its center member is, it'll wear out. This is the primary point in maintaining couplings. There are two main types of couplings: rigid and flexible.

Rigid Couplings

Rigid couplings usually require no further maintenance than correct alignment. Once they're aligned, correct-sized coupling bolts should be installed and then tightened to the correct torque for that

Fin like projections

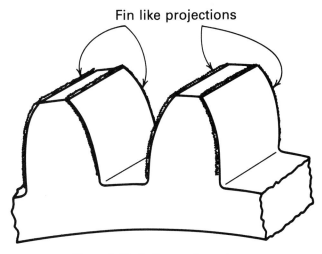

Figure 6-53. Rolling and peening.

bolt. Since there are no moving parts, no wear should occur. On inspection, the bolts should be checked to ensure that they haven't loosened.

Flexible Couplings

Flexible couplings require more maintenance than do rigid couplings. They should be aligned to the same standards, as misalignment causes rapid wear. Flexible couplings can be divided into two classes: mechanical and material.

Cracks

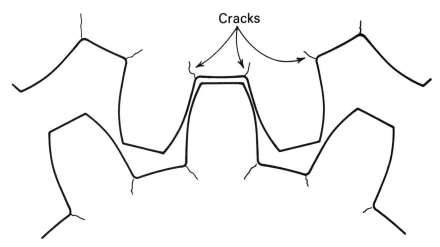

Figure 6-54. Overload cracks in gear teeth.

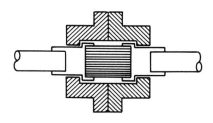

Flexible Grid

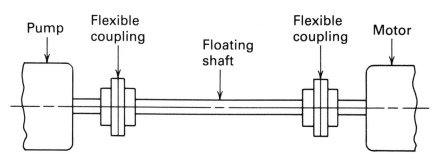

Spindle

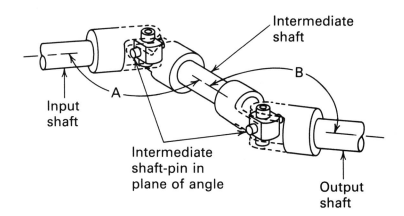

Universal joint

Figure 6-55. Mechanical flexible couplings.

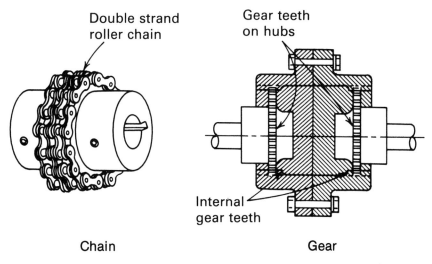

Double strand roller chain

Gear teeth on hubs

Internal gear teeth

Chain   Gear

Figure 6-55. (Continued.)

Mechanical flexible couplings depend on some form of a mechanically flexible element. In this class falls the gear, chain, grid, spindle, and universal joint (Figure 6-55). These all require the use of lubrication to prevent wear. If they aren't lubricated, they'll wear very quickly. The lubricant should be clean. If it isn't, rapid wear will occur as the contamination works through the coupling. If sufficient lubrication is not applied, metal-to-metal contact will occur between contacting metal parts under load, and rapid wear will occur. Care must be taken to ensure that the couplings are lubricated with the proper amount of clean lubricant.

Being packed with too much lubricant can cause problems. The forces in the coupling will churn the lubricant, overheating it and destroying its lubricating ability. This will cause failure of the coupling.

The following guidelines should be used:

gear — half full of clean lubricant
chain — packed with clean oil
grid — packed with clean grease
spindle — same as gear
universal joint — dependent on application.

Material flexible couplings use some form of flexible material between the two coupling halves to absorb some limited misalignment

(Figure 6-56) but this isn't a cure-all. The inspector should give attention to the flexible member during inspection. If the member has been hot or is cracked and showing wear, it should be replaced and the alignment checked. When the alignment is bad, it flexes the material, heats up, and wears more rapidly than it should.

One final point that should be considered with any coupling is the condition of the equipment the coupling is mounted on. The base should be checked and the base bolts should be checked for tightness. Bad bearings in the drive and loose base bolts will greatly accelerate wear of the coupling.

Proper attention should always be given to details when dealing with couplings.

Brakes

All mechanical equipment has some form of brake for stopping or holding. Brake inspections can be divided into two broad categories: mechanical and electrical.

Mechanical inspections include examination of the adjustment to ensure that it's correct. Poor adjustment of brakes can lead to other problems. As the adjustment is checked, the condition of the linings should be inspected. The linings should always be replaced before the backing of the shoes can come into contact with the drum or rotor (Figure 6-57). After this is accomplished, the linkage of the brake should be checked. All bushings, bearings, and mechanical arms should be checked for wear or looseness (Figure 6-58). Any excess play will not allow the brake to function correctly.

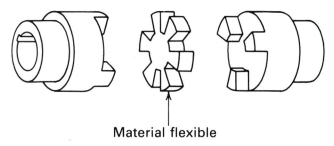

Material flexible

Figure 6-56. Flexible coupling.

Brake linings

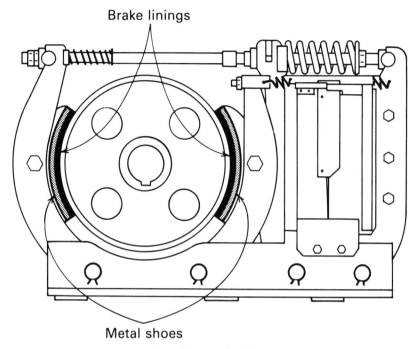

Metal shoes

Figure 6-57. Brake linings.

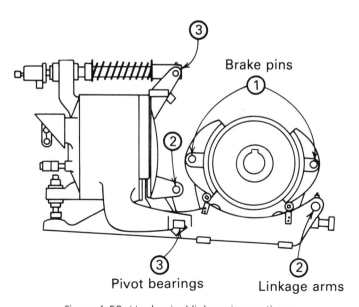

③

Brake pins

①

②

③

Pivot bearings

②

Linkage arms

Figure 6-58. Mechanical linkage inspections.

After all mechanical adjustments and inspections are made, the electrical part of the brake should be checked. The control system should be checked for any worn contacts or relays. All wiring should be checked to ensure that there's no broken insulation and that the wiring is held firmly in place and not loose. The brake coil may be checked to ensure that it isn't grounded or shorted.

Brake inspection varies from make to make, but these points should help pinpoint any problem areas before they allow a breakdown.

7 Fluid Power Systems

Fluid power systems are growing in popularity in industry today. Any effective preventive maintenance program must include inspection points for the various components in the fluid power system, which include the following:

1. pumps or compressors
2. fluids
3. valves
4. hoses
5. actuators.

Pumps

There are two types of problems occurring with pumps: noise and heat.

These two conditions indicate to the inspector problems that can have a variety of causes. Let's examine the causes separately.

Noise

Pump Running Backwards: Noise in a fluid power system is an indication that wear is taking place. One of the possible reasons for noise in a hydraulic system is that the pump is running backwards (Figure 7-1). While this is not the most common cause of noise, it is one that should be considered, especially if the noise begins to occur after the pump or motor has been replaced. Some hydraulic pumps are reversible while others are not. If the pump is not reversible, and the components were assembled so that the pump runs backward, then the internal components will operate in a manner they weren't designed to. This will cause rapid wear and noise. The inspector should be alert to this fact.

Air Leaks: Another more common cause of noise in a hydraulic system is an air leak on the inlet side of the pump (Figure 7-2). This allows the air to be trapped in the oil. As the air passes through the pump, it's subjected to extreme pressure, causing the oil to dissolve into the fluid with a loud noise, which is transmitted throughout the piping system.

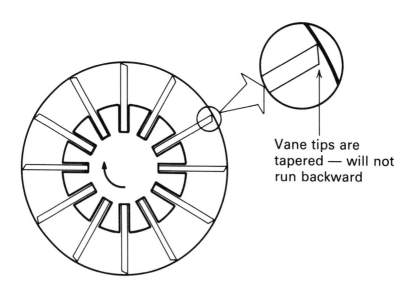

Vane tips are
tapered — will not
run backward

Figure 7-1. Improper vane rotation

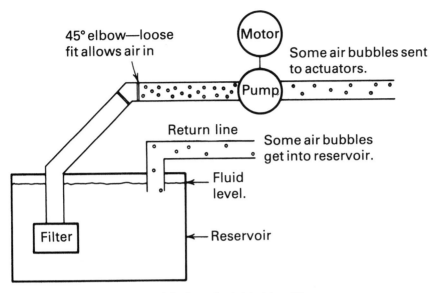

Figure 7-2. Air leak on the inlet side of the pump.

What are some of the ways that air can get into the inlet? One common way is through a crack in the piping or a loose joint in the inlet piping system (Figure 7-3). The pressure in the inlet piping is less than the atmospheric pressure outside, and the air is forced into the line. This will require close scrutiny on the part of the inspector in order to find the inlet air leak.

Another way for the air to enter the system is through the pump seal (Figure 7-4). This is a common occurrence especially in older installations. It's easier to check this problem. Take a small container of oil and pour it around the pump seal as it's running. If the loud noise stops, then you have found the leak. The oil will seal the shaft preventing the entrance of the air into the system. It should quiet down almost immediately. Taking the pump off line and replacing the seal or changing the pump is the recommended cure.

Air can enter the system when the fluid level becomes too low (Figure 7-5). This allows the air to enter the inlet line while it's in the reservoir. The cure is simply to add more fluid until the level is at the recommended point. The air will circulate out of the system after a period of time, come to the reservoir, and bubble out to the atmosphere.

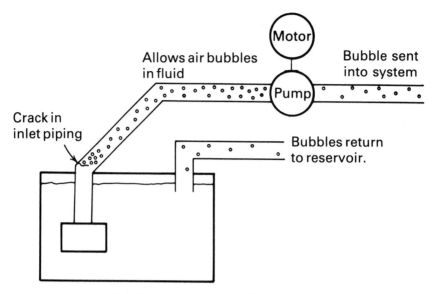

Figure 7-3. Air leak through a crack in the inlet piping.

One point to keep in mind with inlet air leaks is that the fluid always appears milky in the reservoir, due to the trapped air. If noise is present and the fluid is milky or foamy, you probably have an inlet air leak.

Cavitation: What if the pump is making a noise as if it has an air leak, but there's no foam or milky fluid in the reservoir? The problem here is cavitation, which has the same symptoms as an inlet air leak. This is probably one of the most misused terms in fluid power. Cavitation is the term used when the inlet pressure of the system drops below the vapor pressure of the fluid being pumped. All fluids

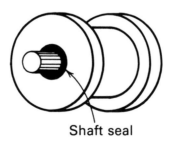

Shaft seal

Figure 7-4. Air leak through the pump seal. When shaft seal wears, air gets into the pump chamber.

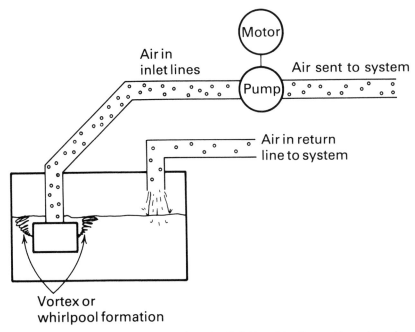

Figure 7-5. Air enters the system when the fluid level is allowed to get too low.

have dissolved air in them. If the inlet pressure drops below a certain point, this dissolved air comes out of the fluid. As it passes through the pump, it goes from an area of low pressure to one of high pressure (Figure 7-6). The air or vapor bubbles collapse (implode) back into the fluid with considerable force. There will actually be enough force to damage the pumping elements. The implosion begins eroding away the surface of the pumping element until failure results — thus the importance of preventing cavitation in a fluid power system.

What causes cavitation? There is a variety of causes, and it's occasionally difficult for the inspector to determine the actual cause.

If it's a new system or a revamped system, the cause can result from a design problem. This is not likely on a system that has been in operation for a time, with no history of cavitation problems. A new system may be forced to lift the fluid from the reservoir to the pump inlet (Figure 7-7). Too much lift will require a pressure below the vapor pressure of the fluid. In this case, engineering specifications should be consulted.

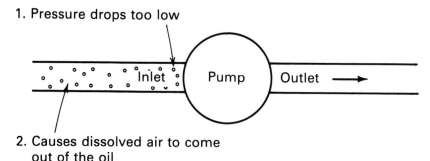

Figure 7-6. When the inlet pressure drops below the vapor pressure of the fluid being pumped, cavitation occurs.

If it's an older system and the problem just appeared, there are several possibilities. One of the most common is a dirty inlet filter. If the filter is dirty enough to restrict flow to the pump, it won't allow the pump to draw enough oil into the intake (Figure 7-8). The pressure at the inlet drops while trying to draw more oil into the pumping chamber. As the pressure drops to the vapor pressure of the fluid, cavitation begins. If the inlet filter is not changed, rapid wear and premature failure of the pump will occur.

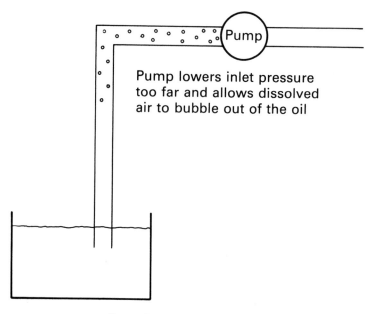

Figure 7-7. Typical cavitation.

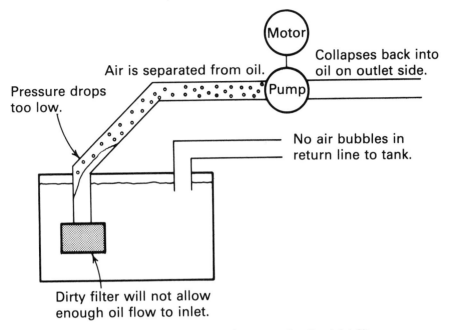

Figure 7-8. Cavitation can occur because of a dirty inlet filter.

Another cause of cavitation is the use of incorrect oil. If the viscosity of the oil is too great, it won't flow into the inlet at the correct rate. It could be compared to trying to draw cold molasses through a straw — it takes considerable effort. In the case of the hydraulic system, it takes more effort to draw the thicker oil into the inlet. This drops the intake pressure below the vapor pressure of the fluid, and cavitation occurs.

Sometimes cavitation occurs only after some component changes in the system. If the inlet line is replaced with one that's smaller in diameter, cavitation will occur. This makes the pump work harder to draw the oil into the inlet (Figure 7-9). If the motor driving the pump is replaced with a motor having a higher RPM, the pump will run faster, trying to pump more oil than the system was designed to. This will decrease the pressure at the inlet to a level that may cause cavitation.

Bearing Problems: Another cause of noise that the inspector should be alert to is bearing problems, which are usually classified into two categories: lubrication and alignment.

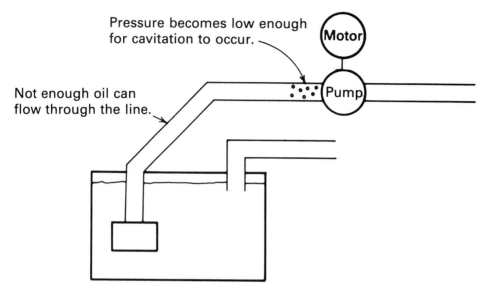

Figure 7-9. If the inlet line is replaced with one that's smaller in diameter, the pressure can drop low enough for cavitation to occur.

Lubrication Problems: Lubrication problems in a pump can be divided into two classes, depending on the type of pump. Some pumps are lubricated by the fluid they are pumping (Figure 7-10), and others (Figure 7-11) are lubricated from an external source (some pumps have sealed bearings and do not require relubrication). The bearing that's lubricated by the fluid that it's pumping is in danger of being contaminated by any dirt in the fluid. Unless a system is carefully and periodically filtered, it's going to contain dirt, which will get into the bearing and cause rapid wear and noise. The fluid's level of contamination should be carefully monitored. If it's contaminated, it's safe to assume that the bearing is also being contaminated. Off-line filtration is recommended to clean the fluid. The cleansed fluid will then flush the contaminants out of the pump and little additional damage will occur. The noise should disappear or diminish at that time.

If the bearing is externally lubricated, there are two dangers. The bearing could once again be exposed to contaminants in the lubricating fluid. Careful filtration should thus be applied to the lubricating fluid. The other problem is that the fluid must be piped into the bearing. If the line becomes clogged, the fluid won't be able to get to the bearing, or gets there only in smaller quantities. This will in-crease the heat and wear in the bearing, resulting in an excessive noise level. Each time the bearing is inspected, it should be observed for correct oil flow.

Alignment Problems: Alignment problems are the other cause of noise in a bearing. If the pump shaft is not correctly aligned with the motor that is driving it, it puts stress on the bearing. The bearing will be excessively loaded on one side, and will wear in such a way that the bearing will have metal-to-metal contact on the loaded side (Fig-ure 7-12). This will result in heat and noise. If it somehow survives for a while under this condition, the seal will usually leak and allow air into the inlet, adding to the noise. Alignment is a must in any pump installation if the pump is expected to last at all. After the pump is aligned, periodic checks should be made to assure it doesn't work loose in operation, causing another misalignment problem.

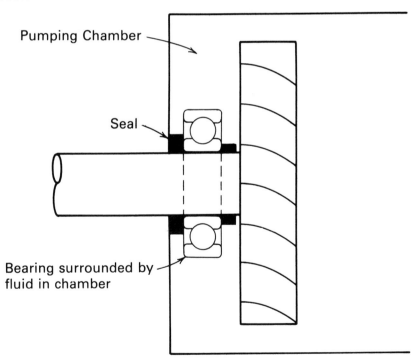

Figure 7-10. Lubrication by the fluid that the system is pumping.

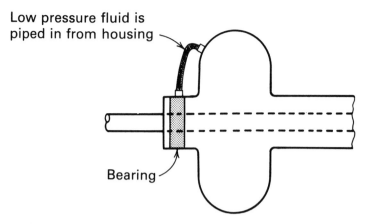

Figure 7-11. Lubrication from an external source.

Heat

Heat in a hydraulic pump can have a variety of causes. It's best to start with the most basic cause — normal wear.

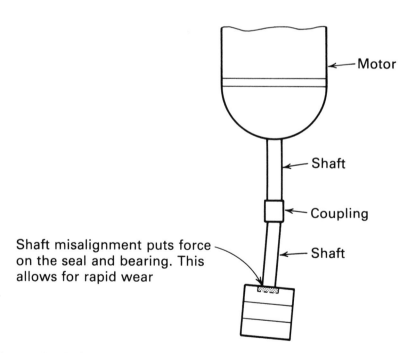

Figure 7-12. Shaft misalignment puts force on the seal and bearing, causing rapid wear.

Worn Components: When the wear in a pump becomes excessive, its efficiency drops. The pump develops enough worn components that it develops internal leaks from the inlet side to the outlet side (Figure 7-13). These leaks are small at first, but the pump must work harder than it was designed to in order to keep the system in operation. As the leaks worsen, more and more fluid leaks past the pump mechanism, making the pump run under pressure at all times. This will generate more heat than the system was designed to dissipate. The heat builds until it becomes excessive enough to damage the fluid (above 140°F). Then the hydraulic system develops multiple problems that take time to diagnose, due to the gums and varnishes building in the oil. It's best for the inspector to find the hot pump as soon as possible to prevent development of further problems.

Pressure Set Too High: If the pressure is set too high in the system, it may require the pump to run under full load for too long a time, generating excessive heat in the system. The pressure of the system should always be an inspection point on any system inspection.

Most hydraulic systems have two valves for the protection of the pump: an **unloading valve** and a **relief valve** (Figure 7-14). The relief valve prevents the pump from developing too much pressure. At a given pressure, the relief valve allows the fluid to pass back to the tank. The unloading valve is set at a pressure less than the relief valve. It has larger ports and piping to allow the pump to pass most

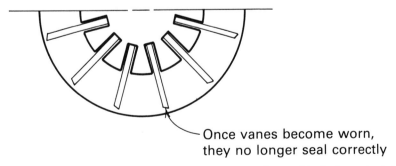

Once vanes become worn, they no longer seal correctly

Figure 7-13. Worn components in a pump can cause internal leaks from the inlet side.

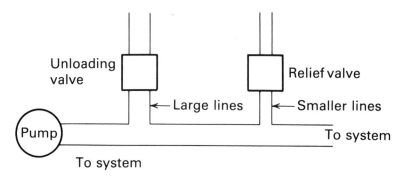

Figure 7-14. Unloading valve and relief valve for protection of the pump.

of its flow across it when the flow is not required by the system. The problem arises when the unloading valve is set above the relief valve. The pump tries to pass all of its flow across the relief valve, which has smaller ports and piping. The flow is passed through under high pressure, keeping a load on the pump and generating excessive heat on the pump and at the relief valve. This situation must be corrected before the system is damaged.

Hydraulic systems set in areas of high temperature, or systems that generate a lot of heat, are equipped with an auxiliary cooler to dissipate the heat. These cooling systems can be air-circulated coolers or water cooled. Whatever method is used, they should be periodically inspected to be sure they are cooling. They may become clogged (Figure 7-15) or perhaps develop a leak and allow water to enter the oil. In either case, the system will suffer. Care should be taken to include all coolers on an inspector's schedule.

Leaks in the Hydraulic System: Leaks in the hydraulic system will also cause the pump to overheat. The pump tries to build pressure but the leak continues to bleed it off. The leak may be an internal leak (keeping the fluid in the system) or an external leak (more obvious because the oil should be visible). Internal leaks can be easy to locate if the line that's hot is traced away from the pump (Figure 7-16). This should be the line that allows the oil flow back to the tank. Attention can then be given to the component that is allowing the flow. External leaks can be traced using the same method. It should end with some visible sign of leakage.

While pumps are the heart of any fluid power system, they need constant monitoring. If trouble is allowed to go unchecked, severe wear will result in the rest of the system. The alert inspector can prevent small problems from becoming major ones.

Compressors

Faulty compressors have almost the same symptoms that a faulty pump can have. Compressor inspection points can be divided into two main classes: heat and noise/vibration.

Heat Problems

Overheating in a compressor can result from the following problems:

1. clogged intake
2. air leaks
3. not unloading
4. cooler problems (multistage).

Clogged Intake: Clogged intake can be caused by a dirty inlet filter or perhaps some foreign material entering the intake (Figure 7-

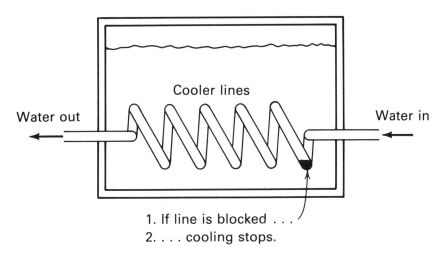

Water out

Water in

Cooler lines

1. If line is blocked . . .
2. . . . . cooling stops.

Figure 7-15. Cooling system with blocked line.

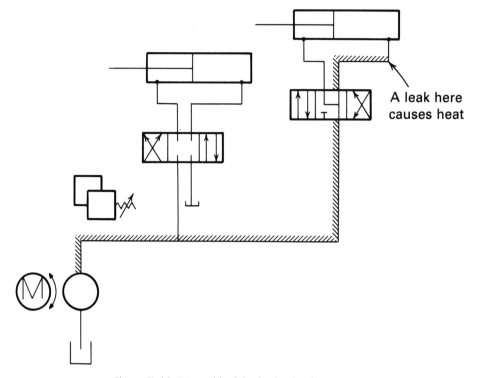

A leak here
causes heat

Figure 7-16. Internal leak in the hydraulic system.

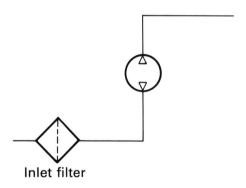

Inlet filter

Figure 7-17. If the inlet filter is dirty, no air can get to the compressor.

17). This reduction on the amount of air available for compression causes the compressor to heat up and run hot. Sometimes the compressor will suck the oil out of its crankcase, due to the low pressure at the inlet (Figure 7-18).

Air Leaks: Air leaks in the system cause the compressor to run all the time trying to maintain the pressure in the lines. This causes the compressor to run hot. This problem also appears in any system where the compressor isn't large enough to supply the needs of all the down line components.

Unloading Problems: A compressor that won't unload tries to maintain constant line pressure. The compressor usually has a check valve, where the compressor is connected to the system (Figure 7-19). The check valve allows the compressor to unload once system pressure is built up and is being maintained. This allows the compressor to stop and cool down. If the check valve is stuck open or there are leaks in the system (as discussed earlier), the compressor must run under continuous load, which will overheat it. All of the above points must be kept in mind while inspecting a hot compressor.

Cooler Problems: If a compressor is multistage, it usually has a cooling device between stages (Figure 7-20). This device lowers the air temperature as it passes from stage to stage. If the cooler isn't working, the hot air from the first stage is compressed in the second

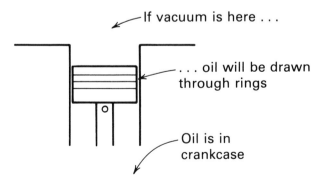

Figure 7-18. Sometimes the compressor will suck oil out of its crankcase because of low pressure at the inlet.

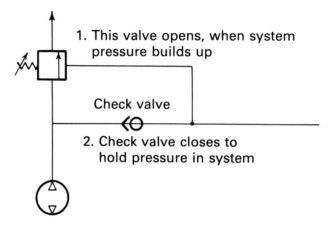

Figure 7-19. Check valve on a compressor. When the system pressure builds up, the valve opens, then the check valve closes to hold pressure in the system.

stage making it hotter yet. This causes the compressor to overheat, lowering its efficiency.

Noise/Vibration

The most common cause of vibration or noise in a compressor is wear. Among the most common components that wear in a compressor are the bearings. Bearing wear begins as vibration, and as it progresses it becomes noisy. As it nears failure, heat also develops. The unit should be taken out of service at this time.

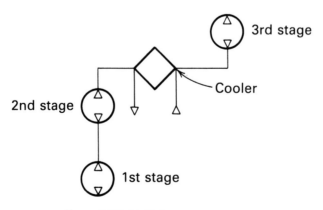

Figure 7-20. Multistage compressor.

Worn rings and mechanical linkage (Figure 7-21) in a compressor also cause noise and vibration. As the components wear, the efficiency of the compressor begins to decrease. An obvious sign is a drop in air flow rate or the inability of the compressor to develop pressure. When these signs begin to occur, the compressor should be scheduled for an overhaul or rebuild.

Since a compressor uses air instead of oil or liquid as a pump, it needs lubrication. The lubrication level in a compressor must be maintained if it's to run efficiently. Low lube levels will result in all three signs of trouble: heat, vibration, and noise.

Some compressors have adjustable linkages, which should be inspected. If these linkages become worn or loose, they'll also cause noise and vibration.

Fluid Inspections

In fluid power systems, clean fluids are an absolute necessity. Care should be taken never to contaminate the fluid. Inspectors should be alert to the following areas of possible fluid contamination.

1. **Storage.** Any fluid in storage has the possibility of becoming contaminated. Fluids should be stored indoors to prevent condensation inside of the container. As temperature changes occur outdoors, the water vapor in the drum may condense, putting water into the oil. This won't be noticed until it accumulates in the system. Any fluid that's put into a hydraulic system should be run through a filter as it's installed. It will contain enough particles of

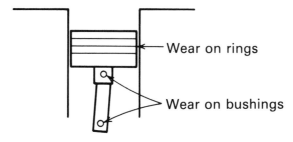

Figure 7-21. Worn rings and mechanical linkage cause noise and vibration.

contamination to harm the system, due to the packing and dirt in the drums.

2. **Adding oil to systems.** Funnels and hoses used to add the fluid should be kept clean. If the hoses, funnels, and containers are not kept clean, dirt will enter the system.

3. **Operation.** The inspector should ensure that all filler caps are in place. All clean-up holes should be closed and sealed. All breather caps should be in place and securely fastened (Figure 7-22). These measures will prevent foreign matter from entering the system.

Accumulators

Accumulators are used in a hydraulic system to store the fluid under pressure. This will enable the system to have a reservoir of stored fluid when running at peak demand. There are several types of accumulators, but the principle is the same: the fluid is stored under pressure. This presents a safety problem when working on the system. If the system is shut off, the accumulator will hold enough fluid under pressure to operate the system for a short period of time. So before working on the system, the accumulator must be shut off or have its pressure bled off.

The most common type of accumulator is one that uses a compressed gas, usually nitrogen, to charge the accumulator. The gas is charged into the accumulator under pressure. The higher the pressure the gas is charged under, the less volume of oil that the accu-

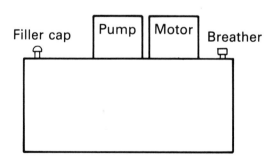

Figure 7-22. Breather caps and filler caps should be in place and fastened to prevent fluid contamination.

mulator can contain. The charged pressure of the accumulator must never be as high as system pressure, and rarely exceeds two-thirds of the system pressure.

One of the most common problems with accumulators is losing the nitrogen charge. When this occurs, the accumulator fills with oil and loses its effectiveness. There must be a given volume of compressible gas in order for the accumulator to function properly. If the gas leaks out, then the pump will run loaded, continuously trying to supply the entire system under high load. This will result in heating and overload on the pump. The solution is to repair the leak and recharge the accumulator to the proper level.

Control Valves

Control valves in fluid power systems have the following problems:

- sticking or binding
- incorrect installation
- coil problems (solenoids)
- water (pneumatic systems).

Sticking or binding valves in a fluid power system are difficult to inspect. If the problem is reported, the inspector should be aware that it's usually dirt or gums (varnishes) that have built up in the system. The average solenoid valve requires 10 pounds of force to shift. If dirt and gums accumulate, it may take as much as 30 pounds, which will burn the coil out (Figure 7-23). This should emphasize the importance of keeping dirt out of the system.

Incorrect installation usually occurs only after an overhaul or replacement of a component in the system. It simply means that someone installed the valve lines incorrectly and the system won't function properly.

Coil problems occur in solenoid valves. Three common causes (aside from the above) are as follows.

1. **High temperature.** When the environmental temperature is above 100°F (Figure 7-24), the resistance in the coil increases, which reduces the amount of current. This means the coil won't

Dirt in these areas may
stop spool from shifting

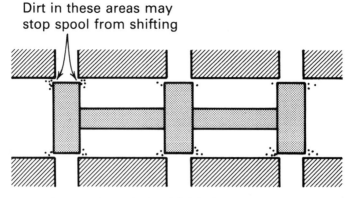

Figure 7-23. Contamination blocking spool ports.

have as much shifting force. It can't close and the coil will burn up. It's important to watch the surrounding temperatures.

2. **Wear.** Wear on the armature causes excessive clearance (Figure 7-25). This means more hammering and forceful closings, resulting in more rapid wear and eventual failure.

3. **Chattering.** If the wiring to the solenoid is found to be good, and the power supply is the correct voltage, the problem may be simple. The armature fits in the coil only one way. If it's turned around during repair or troubleshooting, chattering may result.

Water can cause severe problems in pneumatic valves (Figure 7-26). If water is allowed to be in the valve during a down period, it

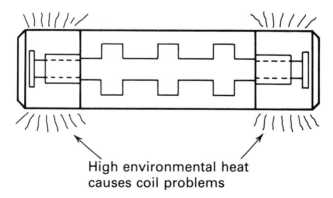

High environmental heat
causes coil problems

Figure 7-24. Directional control valve.

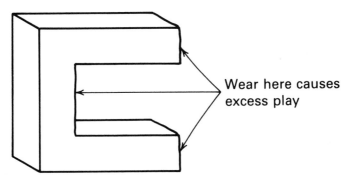

Wear here causes
excess play

Figure 7-25. Wear on the armature.

may cause enough rust and corrosion to freeze the valve in place. It's important that all drains and traps be periodically inspected to prevent the buildup of water in a pneumatic system. Lubricators are also important to provide the correct lubrication.

Hoses

Almost any plant will use some hoses in its fluid power systems. Hoses have definite advantages over fixed piping in some applications. They also have some problems. Some inspection points to consider are:

- exterior damage
- correct length
- rotation.

Exterior damage to a hose usually results from some form of abrasion. An obstruction rubs the hose until the hose has a hole worn in it (Figure 7-27). The best way to avoid abrasion damage is to be sure the hose is clamped in place securely, and not free to move around. If there's an obstruction rubbing against the hose, it's best to shield it with an abrasion guard, which is designed to eliminate rubbing on the hose by absorbing the rubbing action itself.

Using the correct length of hose is also important. It can't be too long because of the possible damage to the hose. It can't be too short or it will put unnecessary stress on the fittings (Figure 7-28). A hose under pressure pulsations can change length from -2 to +4%.

Most pneumatic systems
have vertical drops

This allows water to accumulate,
which will get into valves

Figure 7-26. Water in pneumatic valves can cause severe problems.

Careful sizing is a must. If the hose must be fastened to a carriage or trolley that will be moving a distance, it's best to use some form of hose reel or a festoon system to prevent unnecessary wear.

Rotational wear occurs when the fitting that the hose is fastened to must rotate. This sets up rotational forces in the hose that put stresses on the connections. This results in wear at this point and failure of the fitting (Figure 7-29). It's better in this situation to use a rotational fitting, which does the rotating and leaves the hose free of any rotating forces.

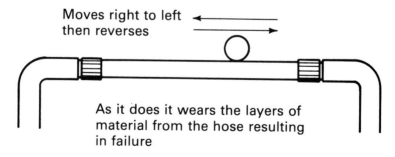

Moves right to left
then reverses

As it does it wears the layers of
material from the hose resulting
in failure

Figure 7-27. Abrasive wear on hoses.

Actuators

Actuators can take the form of cylinders or motors that are powered by fluid power systems. These devices have points that inspectors need to inspect to ensure proper operation. The problem areas are as follows.

1. **Erratic action** is usually caused by some fluctuation in the flow of fluid to the actuator. If the hydraulic systems get air in the lines, the inherent compressibility of the air will make the actuators erratic.

2. **Sticking or binding** can be caused by misalignment of the actuator, dirt in the actuator, varnish or gums building up in the actuator, or simply worn parts.

3. **Valve problems** were discussed above, but they can cause problems with the actuators, so they should be given consideration during any inspection.

4. **Leakage in an actuator** can be internal or external. Internal leakage will give an external sign of heat (Figure 7-30). If it's suspected, the inspector can remove the return line; if oil is leaking through the actuator, it will leak with the actuator at rest. External leakage is visible at the actuator. It may be due to leaky seals or

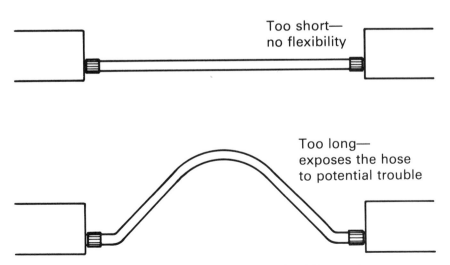

Too short—
no flexibility

Too long—
exposes the hose
to potential trouble

Figure 7-28. The correct length hose should be used.

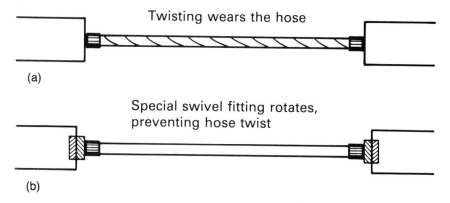

Figure 7-29. (a) Rotational forces put stress on the hose connections. (b) Rotational fittings prevent the hose from twisting.

perhaps loose fittings. Whatever the cause, they should be stopped as quickly as possible.

Pneumatic actuators can be damaged by wet, dirty air. The air must be filtered and dried before it gets to the actuator if damage is to be prevented. Lubrication is also important in a pneumatic system. The lubricant provided by the in-line oilers is the only way the actuators can receive lubrication. If this lubricant is not carefully administered, rapid wear and failure occurs.

Poor hosing and piping restricts flow to the pneumatic actuators causing sluggish action or failure to operate at all. Pneumatic piping is usually lighter duty than hydraulic piping, and is thus more susceptible to damage.

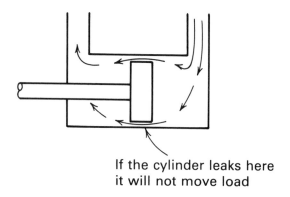

If the cylinder leaks here it will not move load

Figure 7-30. Leakage in the actuator.

A well-engineered fluid power system has tremendous flexibility when it comes to design. If it's to continue to function as designed, it'll need to receive the care that a good preventive maintenance program can provide. The above inspections will enable the inspector to keep the system operating at the optimum level.

8 Electrical Inspections

No preventive maintenance inspection would be complete without including the electrical systems. While mechanical and fluid power systems are prevalent in industry, electrical systems are an absolute necessity. Electrical inspections can be divided into three main areas: 1) the feed lines; 2) switches, relays, and contactors; and 3) motors and generators. Each of these three main divisions has many points to consider.

Incoming Lines

The incoming lines are the large feed lines that supply power to the entire system. These lines are usually large, but the same principles apply to any lines that carry electrical current. There are two main inspection divisions that may be considered: insulation and loading.

Insulation

Insulation of the lines is important because it prevents contact between two hot lines or between a hot line and a ground. The wires usually have an insulating material surrounding them. They may be further insulated by being fastened to an insulator that protects against accidental grounding (Figure 8-1).

The main inspection point of insulation is wear. If the line is attached to a moving piece of equipment, care must be taken to protect the line from abrasive wear (Figure 8-2). If it rubs against any object, the potential is there for abrasive wear. As this wear progresses, the insulation is worn through. Once the voltage finds a path to ground, a short will result causing potentially severe damage to the circuit. This should be a primary inspection point in any electrical inspection.

Another form of wear on the insulation is from environmental conditions. If the insulation is exposed to heat, oil, chemical fumes, or dirt, the potential for insulation breakdown exists. The insulating material is usually susceptible to this type of damage (unless it's a special material made to resist these contaminants). If the insulation is exposed to excess heat, it will harden and develop cracks. The material becomes brittle and will chip away with only slight move-

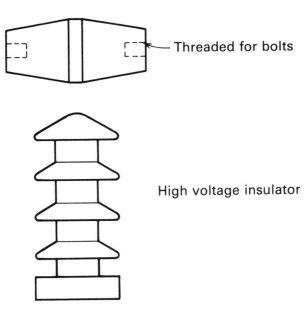

Threaded for bolts

High voltage insulator

Figure 8-1. Incoming lines should be insulated.

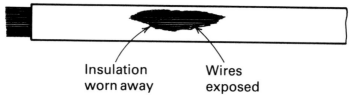

Insulation   Wires
worn away   exposed

Figure 8-2. Abrasive wear.

ment (Figure 8-3). Oil and chemical fumes cause the insulation to soften and swell, causing a split in the insulation. These are both signs of potential trouble that an inspector should be on the watch for.

The insulators that the incoming leads are fastened to are potential problems if they aren't kept clean. Any dirt or contamination becomes a potential path to ground. If a contaminant builds up such a path, a short will occur. Damage results with a shutdown of equipment quite likely afterward. Insulators should be periodically blown off with air or cleaned with a solvent (after the lines are disconnected from the power source). After they've been in service for a time, it's possible that the insulator may become cracked due to temperature change, vibration, or unintentional damage. Any time an insulator is observed to be cracked, it should be replaced or repaired to prevent possible grounding or shorting of the line.

Another place to inspect the terminals (Figure 8-4) is the point where the lines come in. If the terminals become loosened, excessive heat will develop. This is usually evident by the discoloration of the terminals. If this condition is observed, the power should be removed and the terminals tightened as soon as possible. If corrective action isn't taken, the intense heat will eventually destroy the terminal connector and possibly do enough damage to take the unit off line until new terminals can be installed.

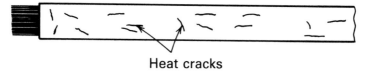

Heat cracks

Figure 8-3. Heat cracks caused by excessive heat.

Loading

Loading on incoming lines is an important item to inspect. If the lines become overloaded, they'll heat up, and damage to the lines and the insulation will result. Loading on the lines can be checked by a portable ammeter called a tong meter (Figure 8-5). This device is extremely handy for the inspector because the line doesn't have to be disconnected to check it. The meter is merely placed around the lead, and the scale will read the amperage in the line. This will give the inspector an indication if further inspection is necessary, or if the system is within allowable limits.

Switches, Relays, and Contactors

Once the power is brought into a terminal on a panel board, it usually goes into the control part of the equipment. Through a series of control relays, the power is distributed to the necessary components on the line. The three rules for preventing most problems with the relay components are: keep it tight, keep it clean, and keep it lubricated. While these three rules keep most components in operation, the inspector needs to watch for the following problems.

Chattering of the Relays

An inspector should watch for a relay that chatters. This is most common on ac controls. This noise tells the inspector to look for

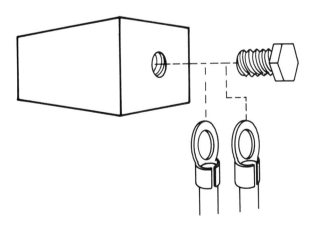

Figure 8-4. Terminals.

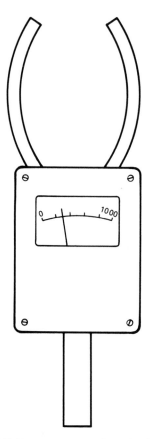

Figure 8-5. Tong meter used to check loads.

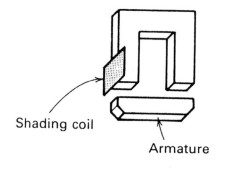

Shading coil

Armature

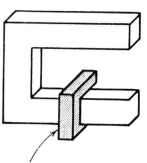

Shading coil has current
induced to help hold
the armature in

Figure 8-6. Shading coil.

three main problems. The first is a **broken shading coil** (Figure 8-6), which is built into the relay to keep it pulled in when the ac current passes through the zero plain, as it does 60 times a second. If this coil is defective, the relay tries to drop out. It pulls back in one-sixtieth of a second, so it chatters in and out. An inspector should note this on the report before excessive wear occurs.

A **dirty magnet face,** even with small amounts of dirt, may cause the relay to chatter (Figure 8-7). This is why the equipment relay should be cleaned periodically with an air spray and electrical solvents — something the inspector recommends as soon as it is noticed that the equipment is becoming dirty.

The third possible cause of chatter in a relay is **dirty contact tips** on the control that holds it in (Figure 8-8). If these tips become dirty, they won't pass the necessary current to hold the relay in. This will allow the relay to drop out, pick up, drop out, pick up, and so forth, and may occur at such a rapid rate that the relay chatters. Again, inspectors should take note of the amount of dirt on the relays.

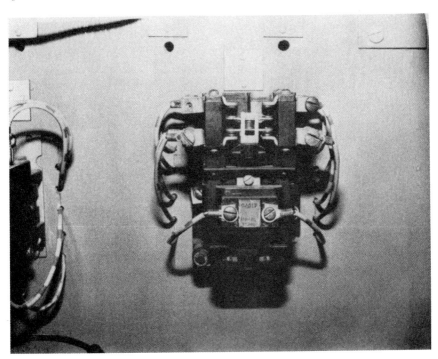

Figure 8-7. Relay magnet face.

Inspect for dirt

Figure 8-8. Auxiliary contact tips.

Heat

A second main point to inspect on relays or contactors is signs of heat. A keen sense of smell is useful here for there is no smell like that of hot or burning electrical insulation. While the contacts themselves won't usually burst into flames, the heat is transmitted along the wire, and the insulation sometimes becomes hot enough to melt or burn.

Discoloration of the contact tips on the relay or contactor points to overheating. Overheated or discolored contact tips are usually caused by three main problems. An **overload in the system** will cause high current to flow through the tips, overheating them. This problem can be checked by the use of the portable ammeter (tong meter). The high current can be read and traced down to its cause.

Weak springs on a relay or contactor don't give sufficient hold in pressure on the tips (Figure 8-9). This has the same effect as a loose connection. The tips will overheat, causing damage to the relay. The inspector should be sure the relays and contactors have the spring recommended by the manufacturer. Either a weaker spring or a stronger spring will cause problems.

Dirt and normal wear are also causes of heat. Dirt causes improper contact between the tips, leading to high current and causing overheating. Normal wear causes the tips to become smaller (Figure 8-10). The smaller they become, the less tension the spring can apply, making the connection loose and causing heat buildup.

Arc Shields

One further point that should be considered at this time is arc shields (Figure 8-11). Arc shields are not found on all contactors, but they are there for the purpose of dampening the arc that is created when the contactor is opened under load. This is like pulling the wire off a spark plug while your car engine is running. It usually draws quite an arc. The same thing happens when a contactor opens under load. To prevent the tips from burning, the arc shield helps control the arc. Arc shields should always be kept clean. If dirt is allowed to build up, the shields may flash over, then they're of no

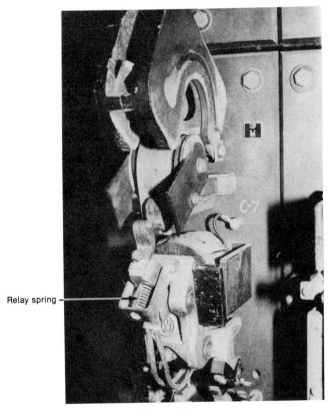

Relay spring

Figure 8-9.

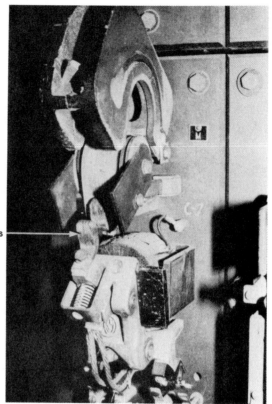

Worn contact tips

Figure 8-10.

use in dampening the arc. They should be checked inside for carbon buildup or copper deposits, which conduct current. If they're kept clean they'll prolong tip life. Alignment of the arc shield (Figure 8-12) is important because they are close fitting. If they become misaligned, they can prevent the contact from closing or can hold it in when it should drop out. This should always be included as an inspection point.

Mechanical Wear

Inspection of a contactor should also include a check for mechanical wear (Figure 8-13). The main inspection points for mechanical wear are:

1. bushings or bearings
2. stops

3. interlocks
4. springs.

Bushings or Bearings: Because the bushings or bearings are the pivot points in the contactors, they are points of wear. If the contactor operates in a dirty environment, these points can wear quickly. The best way to inspect them is to try to move the contactor arm back and forth on the pin. If there's much play, it would be best to schedule a replacement of the bushings. In addition to causing excessive play, the wear of the bushing or bearing will cause the contactor to hang up, not make proper contact, or operate erratically. While the inspector is checking this point, it's also good to check the pivot pins to ensure that they're in place and not loose or free to move.

Contactor Stops: The contactor stops are there to keep the contactor in proper position to pull in when the coil is energized. If the stops wear (and they will after so many operational cycles), the air gap increases in size (Figure 8-14). It will now take more current to

Arc shield

Figure 8-11. Arc shield.

Figure 8-12. Arc shield alignment points.

pull the contactor in, making operation slower. At this point, excess heat is generated in the coil, causing more rapid wear. The inspector should be aware of the normal position of the contactors being inspected and note any that have excessive air gaps.

Interlocks: Mechanical interlocks that reverse direction of a motor or of some component (Figure 8-15) are usually found on relays or contactors. They're usually installed to keep the two contactors from being energized at the same time. They have an adjustable mechanical linkage that holds one of the contactors open if the other one is closed. If this linkage is out of adjustment, they both may be able to pull in at the same time, damaging some component. The inspector should always check to see that they're properly adjusted and in place.

Springs: Contactor spring tension is important to prevent a loose connection on their tips. If the spring is broken or replaced with the wrong size, the contactor won't operate properly. A spring that's discolored by overheating is another sure sign that it should be replaced. The heat changes the composition of the spring enough to affect its tension.

Tightness of Connections

One final inspection area of the relays and contactors is the tightness of all connections — mechanical and electrical.

Loose mechanical connections include all supports and fasteners. The contactors (Figure 8-16) are held in place by bolts and must be in certain places if interlocks and wiring diagrams are to be operational. The inspector should be alert for any loose or dangling components. They should be cared for immediately.

1. Bushings
2. Stops
3. Interlocks
4. Springs

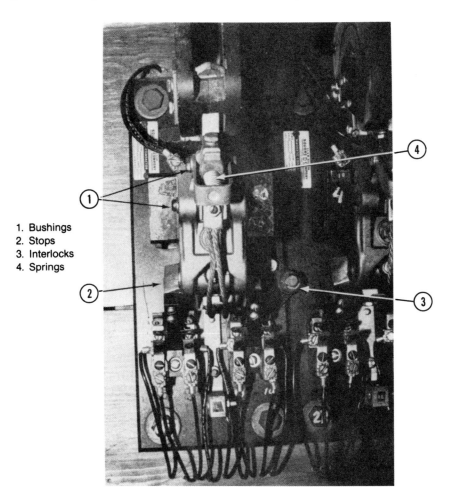

Figure 8-13. Four contactor inspection areas.

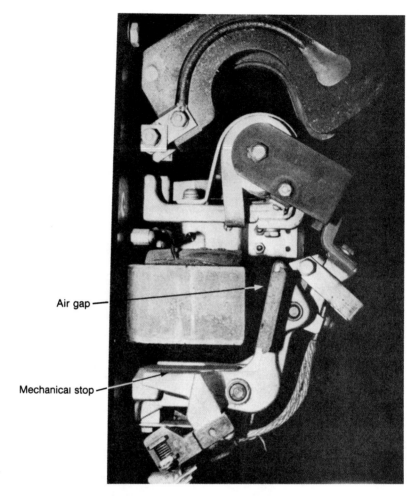

Figure 8-14. Mechanical stop and air gap.

Loose electrical connections cause high current, which results in high heat. This high heat leads to problems if not corrected as soon as possible. The inspector should be alert to any connection that appears discolored, for it will usually be loose.

Another inspection point at this time is the flexible shunts (Figure 8-17). These are strands of wire that carry current to and from the movable arm on the contactor. After so much motion, the wires begin to break, taking on a frayed appearance. Once they have become frayed about 25%, they should be replaced, for they may fail under full load.

These inspections will help keep the control components in good operating condition. This is important for they can cause considerable damage to the motor and drive if they're allowed to malfunction.

Motors and Generators

These two components are considered together because the inspection points are almost identical. The functions of the motor and generator are almost exactly opposite, but they have the same physi-

Figure 8-15. Mechanical interlock.

Figure 8-16. Inspection for loose components.

cal size and shape, and basically the same components. Some points the inspector should be concerned with are:

1. commutator and slip rings
2. bearings
3. vibration
4. wiring and coils
5. overheating.

Commutator and Slip Rings

Commutators (Figure 8-18) and slip rings (Figure 8-19) transmit current between the stationary housing and the rotating part of the motor. Direct current motors and generators use the commutators and brushes to connect the armature to the necessary wiring. The inspector will have to visually inspect the commutator to determine if a problem exists.

The brushes are usually made of carbon and come in different grades or hardnesses. The brushes fit into a holder bolted into the housing. It's important that the brushes be aligned with the commutator. They should also be in a fixed position to avoid excessive sparking. The brush holder should have a space of 1/8 to 3/32 inch of clearance between it and the rotating armature (Figure 8-20). This clearance prevents chipping and rapid wear of the brushes. The brushes are held in the holder by spring tension, usually a bar type that has some adjustment. The tension is determined by the size of the brush, its grade, the speed of the armature, and so forth. It's best

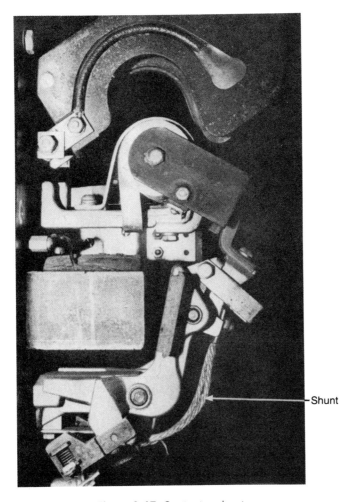

Figure 8-17. Contactor shunt.

Commutator

Figure 8-18. Commutator of a dc machine.

to consult the manufacturer's recommendations for the correct spring pressure, which can be set by using a spring scale. The brushes, brush holders, wiring, and armature should be kept free of all dirt, dust, and oil. If they're not, stray current paths can occur, causing erratic operation or failure of the machine.

Inspection of the armature surface can give an indication of the type of conditions the machine is undergoing during operation.

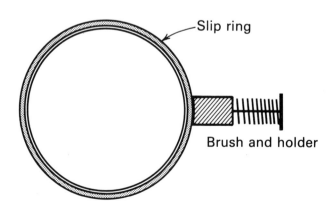

Figure 8-19. Slip ring and brush.

Some of the possible wear patterns and causes are shown in Figure 8-21.

Alternating current machines use slip rings to transmit current. The parts of the ac machine are usually referred to as the rotor (rotating) and the stator (housing). Brushes in the ac machine run on a smooth ring (usually copper). The same material given above on brushes applies to slip rings as far as alignment and grades and pressure. The only problem that rings have occurs when they become worn or pitted. This may cause arcing and burning and rapid brush wear. It's best to smooth them with fine sandpaper if at all possible (not emery cloth). If they're worn too severely, it may be necessary to remove them and cut them on a lathe to give the brush a smooth surface to ride against.

Bearings

Bearings in a motor or generator come in many styles and types. They may require oil lubrication, grease lubrication, or no lubrication (sealed bearings require no lubrication). There are problems with lubrication. It's essential that the bearing receives adequate lubrication. It's just as essential to ensure that it doesn't receive too much, for the excess gets into the motor's wiring and ruins the insulation. If the lubricant is applied with a grease gun through a grease fitting, the bearing can be damaged. As a rule of thumb, a bearing requiring lubrication should be filled from 1/4 to 1/2 full. It's usually recommended that the grease fitting be removed while the machine is in

1/8″ to 3/32″

Figure 8-20. Brush holder clearance.

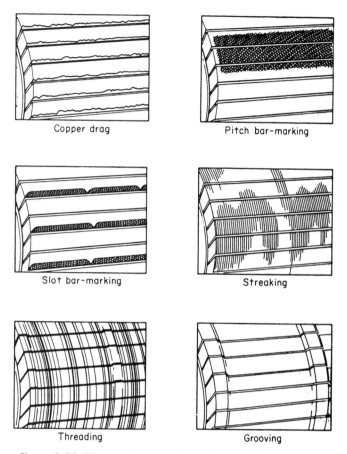

Figure 8-21. Wear patterns on the surfaces of the armature.

operation to allow the excess grease pressure to bleed out. This prevents overgreasing and the subsequent overheating and leaking of the bearing.

Oil-lubricated bearings have oil cups, sight glasses, or oil slingers (Figure 8-22). They should always be checked for correct level to prevent damage to the bearing. Oil slingers rotate with the shaft to splash oil on the bearing while it's in operation. The oil level should always be high enough for the slinger to pick it up.

To check for wear in a bearing on smaller machines, you can try to move the shaft laterally (Figure 8-23). You may be able to get vertical motion as well, depending on the wear. On larger machines, a prying device may give an indication of movement. If this procedure doesn't give the answer, the inspector can just look at the shaft to see if there are clearances around the motor housing. If there is excess clearance at the vertical top of the shaft, wear has occurred. If the worn bearing isn't spotted in time, the wear can become such that the armature or rotor will actually rub into the machine's windings. This will result in costly repairs. The inspector should be alert to this wear.

Vibration

Vibration in a machine is a sign of a problem and can be caused by bad bearings (replacement should be made), misalignment (proper alignment procedures should be followed), or electrical problems with the windings (a megger or tong meter can be used to check). Vibration can also possibly be caused by a balancing problem. Whatever the problem is, it should be diagnosed and solved quickly. Vibration creates stresses in machines that they're not designed to take. If it goes uncorrected, a failure will result.

Wiring and Coils

The wiring and coils in a machine (Figure 8-24) can be checked for shorts and grounds with tong meters or meggers. A tong meter has been mentioned before. A megger, or megohmmeter as it is properly called, is a meter used to determine the insulation resistance of the wiring. This gives the inspector an indication of the condition of the wiring in the machine. The wiring is connected to the megger, and the insulation value is read. A low reading indicates poor insulation and possible problems. A high or good reading indicates the insulation inside the motor would not be a problem. This check can be made without disassembling the machine. The tong meter can be used to ensure that the machine isn't pulling excess current. If the current is high, another consideration might be that the machine is in good condition and that whatever it's connected to

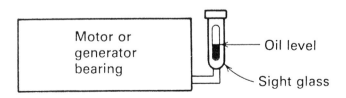

Figure 8-22. Sight glass.

is causing it to pull more of a load. The inspector has to be alert to all possibilities when unusual conditions exist.

Overheating

Overheating in an electrical machine can be caused by many things:

1. clogged air vents
2. defective coolers
3. heaters left on
4. overloads
5. defective internal wiring
6. defective bearings.

Most of these points have been previously discussed in the text, and those that were not are self-explanatory. Clogged air vents should be cleaned, coolers can be checked for proper operation, heaters can be turned off.

Conclusion

The three classes of inspections discussed (mechanical, fluid power, electrical) must be treated equally. A plant that makes inspec-

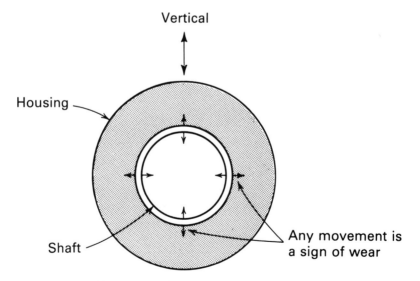

Figure 8-23. Checking for wear in a bearing.

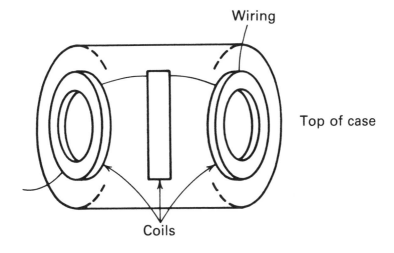

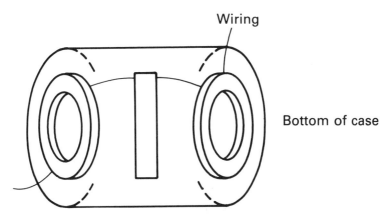

Figure 8-24. Motor wiring diagram.

tions in only one or two groups will have trouble with the third. The idea behind preventive maintenance is to be thorough and complete. It will do little good to spend the money to begin a program, get it in operation, and then not follow up. The bottom line with any PM program is to reduce costs by reducing downtime and production losses.

If the procedures discussed are used as a beginning, a good program will evolve. It would be impossible to describe a program that would apply in all industries in all locations. But by using a basic guide, competent maintenance managers will be able to build a program that will fulfill their individual needs. If industry is to survive in the competitive world market, it will have to utilize a PM program. The challenge is there. Are you up to it?

Appendix A
Troubleshooting
Charts

Table A-1
Pneumatic Troubleshooting

Problem	Possible Cause	Correction
Low air pressure.	1. Compressor volume insufficient.	1. Install larger compressor or install a receiver in the system.
	2. Air leaks in system.	2. Repair leaks.
	3. Internal compressor components worn.	3. Repair or replace worn components.
Noisy operation.	1. Loose or defective components in system.	1. Repair or replace defective components.
	2. Inadequate lubrication.	2. Increase lubrication.
	3. Misalignment.	3. Realign components.
	4. System components loose.	4. Tighten all components to specified torque values.
High air temperature.	1. Compressor intercooler not working.	1. Clean and flush all lines; insure good coolant flow.
Insufficient air volume in system.	1. Plugged inlet filter.	1. Change or clean inlet filter.
	2. Plugged downstream filter.	2. Clean or change fliter.
	3. Lubricator installed backwards.	3. Check and reverse if necessary.
	4. Worn compressor components.	4. Repair or replace necessary components.
	5. Receiver full of condensation.	5. Drain receiver.
Oil in lines.	1. Worn compressor rings.	1. Replace rings.
	2. Defective lubricator.	2. Check and adjust flow rate.
	3. Improper maintenance practices.	3. Use correct methods of injecting oil into the system.

Problem	Possible Cause	Correction
Water in system.	1. Defective separator.	1. Repair separator.
	2. Moisture entering inlet.	2. Reposition inlet.
	3. Cooler not working.	3. Drain and repair cooler.
Sluggish operation of system components.	1. Insufficient air flow.	1. See "Insufficient air volume in system" above.
	2. Water in lines.	2. See "Water in system" above.
	3. Dirt and sludge in the lines and/or in components.	3. Clean separator and filters, check for correct operation of all filters and separators.

Table A-2
Electrical Troubleshooting
Part A: Applies to All Motors in General

Problem	Possible Cause	Correction
System inoperative.	1. No voltage supply.	1. Check all fuses and circuit breakers.
	2. Defective operating of coils.	2. Use a meter to read operating voltage. Any coil that does not operate with line voltage across it will probably be defective.
	3. Control circuit malfunction.	3. Using a blueprint, check for proper circuit operation.
Only part of the circuit works correctly.	1. Control circuit malfunction.	1. Check blueprint for correct operation and sequence.
Relay contactor buzz (A.C.)	1. Broken or defective shading coil.	1. Repair or replace.
	2. Contactor arm out of alignment.	2. Readjust and align contactor.
	3. Dirt between contact surfaces.	3. Clean dirt out of contactor.
	4. Insufficient line voltage.	4. Check and adjust for proper adjustment.
	5. Poor contact between tips energizing the relay coil.	5. Check and adjust the hold in tips.
Overheating of relay tips.	1. Too high current.	1. Reduce load or install larger contactor tips.
	2. Loose contactor tips.	2. Clean and retighten all loose connections.
	3. Weak spring pressure.	3. Adjust or replace defective springs.
	4. Contaminants on relay and/or tips.	4. Clean all affected areas.
	5. Arc shield and/or blowout coil defective.	5. Replace all defective components.

Problem	Possible-Cause	Correction
Welding together of relay tips.	1. Excessive current.	1. Use larger tips or reduce line current.
	2. Incorrect spring tension.	2. Adjust or replace contact springs.
	3. Contactor buzz.	3. See "Relay contactor buzz" above.
	4. Poor tip alignment.	4. Adjust so entire tip-surfaces touch simultaneously.
	5. Movable arm does not close against coil.	5. Check for mechanical bind, or low voltage or the operating coil.
	6. Bouncing of tips.	6. Check for loose components or external vibration.
Short contactor tip life.	1. Excessive current on tips.	1. Check and reduce current or use larger tips.
	2. Abrasives on tips.	2. Clean affected areas and protect from additional contaminants.
	3. Contactor buzz.	3. See "Relay contactor buzz" above.
	4. Low spring pressure.	4. Adjust or replace springs as needed.
	5. Defective arc shields.	5. Repair or replace defective components.
Repeated tripping of overloaded relay.	1. Electrical overload.	1. Find defect in the circuit.
	2. Incorrect overload.	2. Install the correct size overload.
	3. Excessive environmental temperature.	3. Cool relay to acceptable temperature levels.

Problem	Possible Cause	Correction
Failure of relay to pick up.	1. Low line voltage.	1. Increase voltage to correct level.
	2. Damaged coil.	2 Repair or replace coil as necessary.
	3. Excessive air gap.	3. Adjust air gap to correct level.
	4. Mechanical bind.	4. Repair damaged part.
	5. Loose connection.	5. Repair connection.
Failure of relay to open up.	1. Contacts welded together.	1. Replace damaged tips.
	2. Mechanical bind.	2. Repair damaged parts.
	3. Coil fails to de-energize.	3. Check for defect in control circuit.
Hot bearings.	1. Bent armature shaft.	1. Repair or replace damaged shaft.
	2. Excessive overhung loading.	2. Correct the load on the armature shaft.
	3. Misalignment of drive to motor.	3. Correctly align the drive.
	4. Insufficient lubricant.	4. Correct the lube level to recommended amount.
	5. Worn bearing.	5. Check wear and replace bearing if necessary.
	6. Overload on bearing.	6. Check drive to ensure that it is not putting excessive load on bearing.
Motor dirty.	1. Air flow blocked.	1. Clean vents to insure good air flow.
	2. Excessive dirt in environment.	2. Dismantle motor and clean thoroughly. A clean motor runs cooler.

Table A-2
Electrical Troubleshooting
Part B: Applies to D.C. Motors

Problem	Possible Cause	Correction
Motor fails to start.	1. Brush/armature contact is poor.	1. Replace brush springs, and brushes, check for free movement in the brush holder.
	2. Bad bearings.	2. Check for free rotation.
	3. No control circuit.	3. Check for proper power to circuit.
	4. Defective field coils or armature.	4. Repair or replace necessary components.
Motor runs too fast.	1. No load.	1. Add load to motor.
	2. Overvoltage in circuit.	2. Check circuit for correct voltage, adjust as necessary.
	3. Shorted field coil.	3. Check for correct resistance in coils.
	4. Incorrect resistance.	4. Check resistance in entire circuit.
Motor runs away.	1. Incorrect resistance.	1. Check contactor sequence in circuit to insure proper resistance in the circuit; also check for defective resistors.
	2. Shorted field coil.	2. Check resistance of motor coils, replace any defective coils.
Motor runs slow	1. Low voltage.	1. Check circuit for correct voltage, adjust as necessary.
	2. Overload.	2. Check motor for free rotation, check drive for defective components causing overload.
	3. Neutral position of brush holders incorrect.	3. Check manufacturer's position recommendations.

Excessive vibration	1. Armature out of balance.	1. Remove and balance armature.
	2. Misalignment.	2. Check to ensure that the drive and motor are correctly aligned.
	3. Transient from the drive.	3. Check to ensure that the drive is free from vibration.
	4. Loose base bolts or mounting support.	4. Tighten base bolts or correct mounting support.
	5. Mounting base uneven.	5. Shim motor as necessary.
Poor commutation and/or sparking.	1. Brush holders not at true neutral.	1. Consult manufacturer's recommendations for brush settings.
	2. Mica protruding beyond segments.	2. Undercut mica to proper level.
	3. Loose or high bar in armature.	3. Remove armature and have bars tightened.
	4. Wrong grade of brush.	4. Use only the brush grade specified by the manufacturer.
	5. Brushes stuck in holders.	5. Free up or replace brushes, adjust for free movement in holder.
	6. Poor spring pressure.	6. Adjust pressure or replace springs.
	7. Vibration.	7. Correct source of vibration.
	8. Abrasive dust in motor.	8. Clean motor and protect from additional contamination.

Table A-2
Electrical Troubleshooting
Part C: Applies to 3-Phase A.C. Motors

Problem	Possible Cause	Correction
Motor fails to start.	1. Blown fuses.	1. Replace fuses and check for overload in circuit.
	2. Unit single phasing.	2. Check supply for correct feed; also check motor to ensure that windings are open.
	3. Mechanical overload.	3. Reduce overload so that the motor can turn the load and stay within its rating.
	4. Insufficient voltage.	4. Check the supply, adjust the voltage to the correct level.
Motor runs hot.	1. Overload.	1. Reduce the load to acceptable level.
	2. Insufficient voltage.	2. Clean ventilation ducts to allow for free air circulation.
	3. One motor phase may be open.	3. Check the motor to be sure all phases and windings are in good condition.
	4. Incorrect supply voltage.	4. Check supply voltage and adjust to proper level.
	5. Rotor rubbing stator.	5. Check for proper clearance between rotating and stationary components.
Wrong rotation.	1. Incorrect phase connections.	1. Switch any two power leads at the disconnect switch.

Problem	Possible Cause	Correction
Excessive vibration.	1. Poor alignment between motor and the drive.	1. Correctly align motor to the driven unit.
	2. Motor and coupling out of balance.	2. Dynamically balance the motor and coupling.
	3. Defective bearings.	3. Replace necessary bearings.
	4. Excessive shaft runout.	4. Reduce runout by adjustment or replacement of bearings.
Noisy operation.	1. Fan rubbing.	1. Make necessary adjustment to prevent the fan from rubbing.
	2. Loose base bolts.	2. Properly tighten base bolts.
	3. Defective bearings.	3. Replace necessary bearings.
	4. Motor unbalanced.	4. Dynamically balance motor.

Appendix B
Sample
Preventive
Maintenance
Inspections

Appendix B
Sample Preventive Maintenance Inspections

Inspection 1: Typical Motor and Gearcase Combination

Refer to Figure B-1 for this section.

Electric Motor

Item 1 in the figure is the electric motor. The most common inspection on the motor during operation is for heat. The motor should be approximately 20 to 25°F higher in temperature than its surrounding environment. If the motor is higher in temperature than this, its life will be shortened due to the effect the heat has on the insulation. For every additional 20°-temperature rise, the life of the insulation will be cut in half. When the insulation fails, so does the motor. If the temperature of the motor becomes too high, efforts should be made to find the problem, so the motor can be cooled back down.

1A denotes the bearings that support the rotating part of the motor (usually the rotor or the armature). The inspection of the bearings should consider three basic factors: heat, noise, and vibration. These three indicators of trouble may appear singly or in any

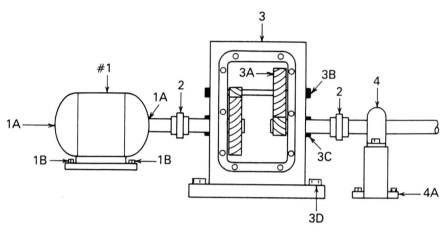

Figure B-1. For inspection 1: typical motor and gearcase.

combination. Monitoring these three conditions may require nothing more than the use of your normal senses. If the equipment is critical in nature, it may require the use of some monitoring or nondestructive testing equipment. For measuring the temperature, some form of hand-held thermometer may be used. If the bearings are in a difficult-to-reach or unsafe location, a permanent monitor may be used. If vibration is to be checked, the use of a hand-held meter may be sufficient. These instruments will be able to indicate such conditions as insufficient lubrication, misalignment, contamination, or normal wear of the bearings. If any of these conditions are detected, measures should be taken to correct the problem as soon as it can be scheduled.

1B denotes the base bolts of the motor. The base bolts should always be checked for the correct torque. If the bolts aren't tightened correctly, they may come loose during the operation of the equipment, resulting in unwanted downtime of the equipment. Also, during removal or installation of the motor, the base should be checked to be sure that it's level and in good shape, to avoid putting additional stress on the motor when it's bolted into place.

Couplings

Item 2 in Figure B-1 is a coupling, which is used to connect two shafts together. The most critical item is the alignment of the couplings. If they aren't aligned to within ±0.005, the coupling will have a shortened service life. While 0.005 is used as a general figure, the closer the alignment is to perfect, the longer the life of the coupling will be. In addition to shortening the life of the coupling, misalignment will also shorten the life of the bearings and related components in the drive. In addition to alignment, certain couplings require lubrication. It is important to provide the correct amount of the correct lubricant to these couplings. If sufficient lubricant isn't provided, metal-to-metal contact inside the coupling will occur, resulting in a reddish brown color to the lubricant and a bluish color to the coupling teeth. Too much lubricant will result in overheating of the coupling due to fluid friction. The extreme heat will then destroy the lubricant, resulting in failure of the coupling.

Gearcase

Item 3 is the gearcase. The gearcase has one important consideration — the lubricant. It must be maintained at the correct level within the gearcase. An insufficient level of lubricant allows metal-to-metal contact between the rotating parts of the gearcase resulting in rapid destruction of all related components within the gearcase.

Item 3A is the gearing. The gears need proper lubrication at all times. If the gears are left to run without lubrication, the tooth surfaces will be destroyed very quickly, resulting in failure of the drive. The tooth surfaces can be inspected periodically for any unusual wear patterns, which may indicate some problem within the case such as contaminated lubricant, misalignment of the gears, insufficient backlash, overload of the gearcase, or pitting of the teeth. Visual inspections should be made semiannually in critical applications, and annually in general applications.

Items 3B and 3C are the bearings supporting the shafts of the gearcase. The most important application with the bearings is the lubricant. The lubrication method may be splash or spray, but it must get to the bearings, or failure of the bearings will result. While this is bad enough, the failure of the bearings can also result in changes of the internal geometries of the gearcase. This will allow damage to the more expensive gears to occur and may result in a more costly and time-consuming breakdown than would have occurred if the bearings had been properly lubricated. It's very important to detect any problems with the shaft bearings in a gearcase before they occur.

3D denotes the basebolts for the gearcase. It's important that these bolts always be tightened to the correct torque values. If the case comes loose during operation, damage will occur to the bearings, gearing, and the couplings. Also, anytime the gearcase is removed, the base should be inspected to ensure that no defect develops that would allow a breakdown during operation.

Bearings

Item #4 is a bearing that supports the end of the drive train. This bearing is very important because it carries the output of the gearcase to its final destination. The inspector should watch for heat, vibration, and noise. If these conditions are detected, it's important to correct the problem before a failure occurs.

4A denotes the mounting or base bolts for the bearing. The same consideration should be given them as was given the base bolts on the motor and gearcase.

Inspection 2: Typical Belt Drive

Refer to Figure B-2 for this section.

Electric Motor

Item 1 is the electric motor. The most common inspection item for the motor is for heat. The motor should be approximately 25 to 29°F higher in temperature than the surrounding environment. If the motor's temperature is higher than this, it will have a shorter life due to the deteriorating effect the heat has on the insulation of the motor. For every 20°-temperature rise above the environmental temperature, the life of the insulation is cut in half. When the insulation fails, so does the motor. If the temperature rises above this level, efforts should be made to find the problem, so the motor can be cooled back down again.

Another inspection point on the motor is the bearings. The bearings support the rotating part of the motor (usually called the arma-

Preventive Maintenance Inspection
Inspection 1: Typical Motor and Gearcase Combination

Check the box or boxes on the right that describe the condition of the units named in the left column.	O.K.	Requires lubrication	Requires adjustment	Requires replacement	Requires cleaning	Excessive vibration	Excessive heat	Loose	See additional comments
1. Electric Motor									
A. Bearings									
B. Base and bolts									
C. Temperature									
D. Vibration									
E. Noise									
2. Couplings									
A. Alignment									
B. Lubrication									
3. Gearcase									
A. Gears									
B. Bearings									
C. Bearings									
D. Base and base bolts									
4. Bearings									
A. Base and bolts									
B. Excessive play or motion									

Additional Comments:

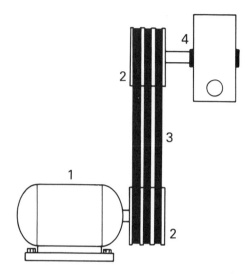

Figure B-2. For inspection 2: typical belt drive.

ture or the rotor). The inspection of the bearings should include three basic items: heat, noise, and vibration. These three indicators of trouble may appear singly or in any combination. Monitoring these three conditions may require nothing more than the use of the inspector's natural senses. If the equipment is critical in the manufacturing process, it may be advisable to use some form of monitoring or nondestructive testing equipment. For monitoring the temperature, some form of hand-held thermometer may be sufficient. If the bearings are in a difficult-to-reach or unsafe area, some form of temperature-monitoring device may be used. Vibration may be measured by a hand-held meter or a permanent monitoring device. Either of these types of vibration meters can determine if conditions such as insufficient lubrication, misalignment, contamination, or normal wear are occurring in the bearing. If any of these conditions are detected, steps should be taken to correct the problem. If they're not corrected, then the bearing will fail, and the replacement will also fail quickly.

Another item on the motor to inspect is the base. The base bolts should always be checked to ensure that they're tightened to the correct torque specifications. If they're not, they may come loose during operation, allowing the motor to misalign with the remaining parts of the drive. Periodically, the base of the motor and the mounting for the motor should be inspected. The foundation may develop faults, allowing the motor to loosen on the base. Also, settling of the base may occur, allowing it to place excessive stress on the motor base. The base should always be completely clean when installing the motor.

Sheaves

Items labeled 2 are the sheaves — the grooves in which the belts run. The contact between the side of the belt and the grooves is the area that transmits the power. It is this area that must be inspected for wear. The sidewall gauge should be used to determine if wear exceeds allowable limits. If it does, the sheaves should be changed to prevent excessive belt wear. The sheaves should also be cleaned periodically with a stiff brush to remove any dirt or deposits that would eventually cause wear on the sides of the sheave. The area where the sheave mounts on the shaft should also be checked to determine if any looseness or excessive wear is occurring. This condition would allow the sheave to slip under heavy loads. If looseness has been observed, corrective action should be taken before a breakdown occurs during operation.

The alignment of the sheaves should also be examined. The sheaves should be in line to prevent excessive wear on the sheave sidewall and the belt. The more exact the alignment is made, the longer the drive life will be. Any time spent aligning the sheaves is time well spent.

Belt

Item 3 in Figure B-2 is the belt. Belt inspections can be merely observations of any unusual wear patterns on the belt. If the belt's surface appears to be normal, with no contamination or cracking apparent, the drive will usually be in relatively good condition.

Heat is a deadly enemy of belt drives. If the belt is exposed to heat, it overcures, causing it to become brittle and crack. Heat can be ambient, or may develop within the drive. Any slippage in the drive causes heat buildup, causing rapid deterioration. This points to the need for proper tension of the belt to eliminate slippage. The best tensioning method is the tension tool, which can be provided by any distributor. Because a belt can slip up to 20% before it makes any noise, sound shouldn't be used to determine if a belt is slipping.

Another test for slippage is the use of a strobe gun. A strobe light "freezes" the belt and sheave during operation. If the belt is slipping, it quickly shows under the light. If the belts exhibit any unusual tracking tendencies, consideration should be given to replacement. This would include running off the sheaves, or turning over in the sheaves. Both of these conditions indicate that some tension members are broken and the belt won't carry its rated load.

Bearings

Item 4 indicates the bearings on the drive. The reason they're included is to point out the effect excessive belt tension or too large a sheave may have on the gearing. If the belt is too tight, it puts an excessive load on the bearings, causing high temperatures, rapid wear, and premature failure. If the drive bearings exhibit these signs, their ratings should be checked, along with the belt tension and the weight of the sheaves.

Inspection 3: Typical Chain Drive

Refer to Figure B-3 for this section.

Electric Motor

Item 1 is the electric motor. The most common inspection item for the motor is for heat. The motor should be approximately 25 to 29°F higher in temperature than the surrounding environment. If the motor's temperature is higher than this, it will have a shorter life due to the deteriorating effect heat has on the insulation of the motor. Every 20°-temperature rise above the environmental temperature cuts the life of the insulation by one half. When the insulation fails, so does the motor. If the temperature rises above this level, efforts should be made to find the problem, so the motor can be cooled back down again.

Another inspection point on the motor is the bearings. The bearings support the rotating part of the motor (usually called the armature or the rotor). The inspection of the bearings should include three basic items: heat, noise, and vibration. These three indicators

Preventive Maintenance Inspection
Inspection 2: Belt Inspection

Check the column that indicates the condition of the unit or what problem exists named in the left column.	O.K.	Requires lubrication	Requires adjustment	Requires replacement	Requires cleaning	Excessive vibration	Excessive heat	Loose	See additional comments
1. Electric Motor									
A. Bearings									
B. Base and bolts									
C. Temperature									
D. Vibration									
E. Noise									
2. Sheave									
A. Sidewall wear									
B. Dirt									
C. Shaft mounting									
D. Alignment									
3. Belt									
A. Cover wear									
B. Tension									
C. Tracking									
4. Bearings									
A. Temperature									
B. Noise									
C. Vibration									
D. Excess play or motion									

Additional Comments:

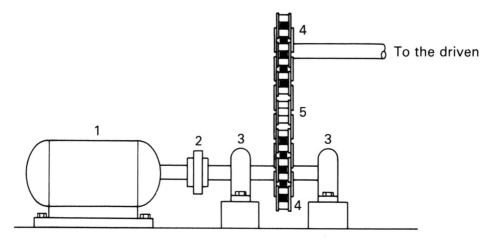

To the driven

Figure B-3. For inspection 3: typical chain drive.

of trouble may appear singly or in any combination. Monitoring these three conditions may require nothing more than the use of the inspector's natural senses. If the equipment is critical in the manufacturing process, it may be advisable to use some form of monitoring or nondestructive testing equipment. For monitoring the temperature, some form of hand-held thermometer may be sufficient. If the equipment is in a difficult-to-reach or unsafe area, some form of temperature-monitoring device may be used. Vibration may be measured by a hand-held meter or a permanent monitoring device. Either of these types of vibration meters will be able to determine if conditions such as insufficient lubrication, misalignment, contamination, or normal wear are occurring in the bearing. If any of these conditions are detected, steps should be taken to correct the problem. If they're not corrected, the bearing will fail, and the replacement will also fail quickly.

Another item on the motor to inspect is the base. The base bolts should always be checked to ensure that they are tightened to the correct torque specifications. If they're not, they may come loose during operation, allowing the motor to misalign with the remaining parts of the drive. Periodically, the base of the motor and the mounting for the motor should be inspected. The foundation may develop faults, allowing the motor to loosen on the base. Also, settling of the base may occur, causing excessive stress on the motor base. The base should always be completely clean when installing the motor.

Couplings

Item 2 in Figure B-3 is the coupling, which is used to connect two shafts together. The most critical item is the alignment of the coupling halves. If they're not aligned within very close tolerances, rapid wear will occur. A rule of thumb is that they must be within 0.005 of an inch. Rigid couplings must be in exact alignment. Flexible couplings may be able to withstand 0.005. The closer the coupling alignment is to being exact, the longer the coupling will last. Misalignment also damages related items in the drive, such as the bearings and shafts. Correct alignment of couplings cannot be overemphasized.

Lubrication is another prime consideration in coupling inspections. If correct lubrication isn't provided for flexible couplings, rapid wear and complete failure will result. If metal-to-metal contact occurs in the coupling, welding and tearing of the coupling material occurs. It's also important not to overlubricate. If overlubrication occurs, fluid friction also occurs, which builds up heat and destroys the lubricant, resulting in rapid wear of the coupling material and failure of the coupling. If a coupling is opened up for inspection and a reddish brown color is observed, it should be cleaned and inspected and then properly relubricated.

Support Bearings

Item 3 refers to support bearings for the sprocket and shaft. These bearings can be inspected by observing the three basic signs of bearing wear: noise, heat, and vibration. If any of these signs are observed, a problem exists with the bearing. This type of bearing in the figure is commonly called a pillow block bearing. If the bearing appears to be in good condition, the lubrication should be checked. If it's at the proper level (covering half the lowest ball or roller if oil-lubricated, or one-third of the housing if grease-lubricated), then it should be safe to move on to the next item.

Chain Sprocket

Item 4 is the chain sprocket. If it's kept in good condition and well lubricated, a sprocket should outlast three or more replacement chains. Sprocket wear occurs on the teeth. If the teeth are becoming hook shaped, then the sprocket will begin rapid wear of the chain.

This will also cause the chain to hang in the sprocket causing overloads on the drive. If the teeth are showing wear on the sides, the sprockets are usually out of alignment. If the alignment isn't corrected, the life of the drive is dramatically shortened.

Chain

Item 5 is the chain. The following inspection points can be applied to any type of chain drive, however, a roller chain drive is considered here. The primary inspection point is lubrication. Since a roller chain wears 300 times faster if it's run unlubricated than when well lubricated, lubrication is a primary concern. The lubricant should penetrate the chain joint to prevent metal-to-metal contact. A rule of thumb is to use a good 30-weight oil at normal temperatures, thinner oil in colder weather, and thicker oil in warmer weather.

The chain should also be inspected to ensure that contaminants aren't clinging to it. If an abrasive material is on the chain, it will accelerate wear between the chain parts and the chain and the sprocket.

If wear is observed on the inside of the chain links, it's a sign that the sprockets are out of alignment. Corrective action should be taken before the chain and sprockets are damaged.

If correct tension isn't kept in the chain drive, wear and shock loading will occur to the chain and the sprockets. As a rule, a 2% deflection at the center of the unsupported span is the correct tension for a roller chain drive. Less than that would cause unnecessary loading on the bearings, any more would cause chain slippage and shock loading on the sprocket.

Inspection 4: Typical Belt Conveyor

Refer to Figure B-4 for this section.

Electric Motor

Item 1 is the electric motor. The most common inspection item for the motor is for heat. The motor should be approximately 25 to 29°F higher in temperature than the surrounding environment. If the motor's temperature is higher than this, it will have a shorter life due

Preventive Maintenance Inspection
Inspection 3: Chain Inspection

Check the column that indicates the condition of the unit or what problem exists named in the left column.	O.K.	Requires lubrication	Requires adjustment	Requires replacement	Requires cleaning	Excessive vibration	Excessive heat	Loose	See additional comments
1. Electric Motor									
A. Bearings									
B. Base and bolts									
C. Temperature									
D. Vibration									
E. Noise									
2. Coupling									
A. Alignment									
B. Lubrication									
3. Bearing									
A. Base and bolts									
B. Excessive play or motion									
4. Sprocket									
A. Tooth wear									
B. Lubrication									
C. Alignment									
D. Mounting to shaft									
5. Chain									
A. Elongation									
B. Side plate wear									
C. Tension									
D. Engagement of sprocket									

Additional Comments:

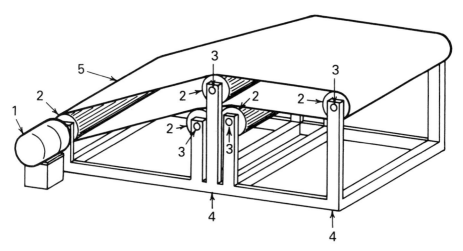

Figure B-4. For inspection 4: typical belt conveyor.

to the deteriorating effect the heat has on the insulation of the motor. Every 20°-temperature rise above the environmental temperature cuts the life of the insulation by one half. When the insulation fails, so does the motor. If the temperature rises above this level, efforts should be made to find the problem so the motor can be cooled back down again.

Another inspection point on the motor is the bearings. The bearings support the rotating part of the motor (usually called the armature or the rotor). The inspection of the bearings should include three basic items: heat, noise, and vibration. These three indicators of trouble may appear singly or in any combination. Monitoring these three conditions may require nothing more than the use of the inspector's natural senses. If the equipment is critical in the manufacturing process, it may be advisable to use some form of monitoring or nondestructive testing equipment. For monitoring the temperature, some form of hand-held thermometer may be sufficient. If the equipment is in a difficult-to-reach or unsafe area, some form of temperature-monitoring device may be used. Vibration may be measured by a hand-held meter or a permanent monitoring device. Either of these types of vibration meters will be able to determine if conditions such as insufficient lubrication, misalignment, contamination, or normal wear are occurring in the bearing. If any of these conditions are detected, then steps should be taken to correct the problem. If they're not corrected, the bearing will fail, and the replacement will also fail quickly.

Another item on the motor to inspect is the base. The base bolts should always be checked to ensure that they're tightened to the correct torque specifications. If they're not, they may come loose during operation, allowing the motor to misalign with the remaining parts of the drive. Periodically, the base of the motor and the mounting for the motor should be inspected. The foundation may develop faults, allowing the motor to loosen on the base. Settling of the base may occur, allowing it to place excessive stress on the motor base. The base should always be completely clean when installing the motor.

Conveyor Roll

Item 2 is the conveyor roll. This is the roll that the conveyor belt is driven by and that it rides on. The main inspection point for these rolls is the surface. It must be in good condition — free of any sharp edges or defects that could cut the conveyor belt. Keep the rolls free of any contaminants that could damage the roll and/or the belt. If contamination buildup is a problem, steps should be taken to remove the contaminants, and methods should be established to keep the contaminants out of this area. The alignment of the rolls is also important. If the rolls aren't kept in alignment with the conveyor framework and tracking system, excessive wear will occur to the roll and to the belt.

Proper tension of the conveyor is a must because slippage of the drive roll causes heat and wear to develop on the roll and the belt. There are many different tensioning devices in a conveyor system. Consult the manufacturer of each conveyor for recommendations on the proper tension.

Bearings

Item 3 is the bearing that supports the conveyor roll. The bearing should allow free rotation of the roll. If the bearing becomes defective and doesn't allow free rotation of the roll, belt wear and wear of the roll occurs. If allowed to continue, the roll may have a flat spot worn on it by the continual rubbing of the belt. Indications of the bearing condition are heat, noise, and vibration of the bearing. If any one of these three conditions is excessive, the bearing is showing signs of wear. Consideration then should be given to the degree of wear and whether this necessitates replacement.

Lubrication of the bearing is also important. Most conveyors have a centralized lubrication system, which is supposed to dispense the correct amount of lubricant to each bearing. If the bearing is continually being replaced, the lubrication system should be checked to ensure that good lubricant flow is possible. The line can become plugged up, restricting the amount of lubricant that the bearing receives.

Framework of the Conveyor

Item 4 is the framework of the conveyor. It should be periodically inspected for broken welds or loose bolts. If weld inspections are

Preventive Maintenance Inspection
Inspection 4: Belt Conveyor

Check the column that indicates the condition of the unit or what problem exists named in the left column.	O.K.	Requires lubrication	Requires adjustment	Requires replacement	Requires cleaning	Excessive vibration	Excessive heat	Loose	See additional comments
1. Electric Motor									
A. Bearings									
B. Base and bolts									
C. Temperature									
D. Vibration									
E. Noise									
2. Conveyor Roll									
A. Surface of roll									
B. Alignment									
C. Contamination									

Inspection 4 continued

Check the column that indicates the condition of the unit or what problem exists named in the left column.	O.K.	Requires lubrication	Requires adjustment	Requires replacement	Requires cleaning	Excessive vibration	Excessive heat	Loose	See additional comments
3. Bearing									
A. Free rotation									
B. Excessive play									
C. Visible damage									
D. Lubrication									
E. Heat									
F. Noise									
G. Vibration									
4. Framework									
A. Loose bolts									
B. Broken welds									
C. Bent or broken supports									
5. Belt									
A. Splice condition									
B. Contamination									
C. Tracking									
D. Tension									

Additional Comments:

made, it's good to use a nondestructive testing method, such as magnetic particle, to assist in finding defective welds. The framework must be kept in good shape. If it isn't, the frame may give or twist enough to misalign the rolls, preventing the belt from tracking correctly.

Conveyor Belt

Item 5 is the conveyor belt. The belt carries the load from one location to another. It should be kept in good condition if it's to work properly. Cuts or tears in the belts should be repaired as soon as possible to prevent additional damage to the belt. There are many different types of belts and belting materials. When repairing or splicing a belt, you should consult the manufacturer's recommendation for methods of repairing the belt.

Contamination should be kept off of the belt as much as possible, but especially on the bottom of the belt where it contacts the pulleys. If contamination is allowed in this area, it results in rapid wear of the belt and pulley.

Belt tracking is important to keep the belt centered on the pulleys. If it's allowed to run to one side or another, the belt will be damaged by the conveyor framework or support.

Inspection 5: Typical Chain Conveyor

Refer to Figure B-5 for this section.

Electric Motor

Item 1 is the electric motor. The most common inspection item for the motor is for heat. The motor should be approximately 25 to 29°F higher in temperature than the surrounding environment. If the motor's temperature is higher than this, it will have a shorter life due to the deteriorating effect heat has on the insulation of the motor. For every 20°-temperature rise above the environmental temperature, the life of the insulation is cut by one half. When the insulation fails, so does the motor. If the temperature rises above this level, efforts should be made to find the problem, so the motor can be cooled back down again.

Another inspection point on the motor is the bearings. The bear-

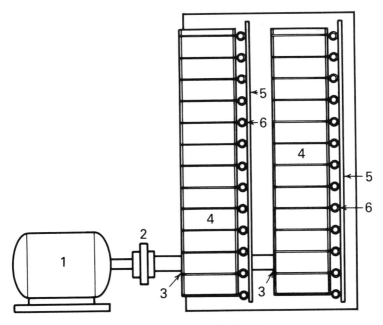

Figure B-5. For inspection 5: typical chain conveyor.

ings support the rotating part of the motor (usually called the armature or the rotor). The inspection of the bearings should include
three basic items: heat, noise, and vibration. These three indicators
of trouble may appear singly or in any combination. Monitoring
these three conditions may require nothing more than the use of the
inspector's natural senses. If the equipment is critical in the manufacturing process, it may be advisable to use some form of monitoring
or nondestructive testing equipment. For monitoring the temperature, some form of hand-held thermometer may be sufficient. If the
equipment is in a difficult-to-reach or unsafe area, some form of
temperature-monitoring device may be used. Vibration may be measured by a hand-held meter or a permanent monitoring device. Either of these types of vibration meters will be able to determine if
conditions such as insufficient lubrication, misalignment, contamination, or normal wear are occurring in the bearing. If any of these
conditions are detected, then steps should be taken to correct the
problem. If they're not corrected, the bearing will fail, and the replacement will also fail quickly.

Another item on the motor to inspect is the base. The base bolts should always be checked to ensure that they're tightened to the correct torque specifications. If they're not, they may come loose during operation, allowing the motor to misalign with the remaining parts of the drive. Periodically, the base of the motor and the mounting for the motor should be inspected. The foundation may develop faults, allowing the motor to loosen on the base. Also, settling of the base may occur, allowing it to place excessive stress on the motor base. The base should always be completely clean when installing the motor.

Coupling

Item 2 in Figure B-5 is the coupling, which is used to connect two shafts together. The most critical item is the alignment of the coupling halves. If they're not aligned within very close tolerances, rapid wear will occur. A rule of thumb is that they must be within 0.005 of an inch. Rigid couplings must be in exact alignment. Flexible couplings may be able to withstand 0.005. The closer the coupling alignment is to being exact, the longer the coupling will last. Misalignment also damages related items in the drive, such as the bearings and shafts. Correct alignment of couplings cannot be overemphasized.

Lubrication is another prime consideration in coupling inspections. If correct lubrication isn't provided for flexible couplings, rapid wear and complete failure will result. If metal-to-metal contact occurs in the coupling, welding and tearing of the coupling material occurs. It's also important not to overlubricate. If overlubrication occurs, fluid friction occurs, which builds up heat and destroys the lubricant, resulting in rapid wear of the coupling material and failure of the coupling. If a coupling is opened up for inspection and a reddish brown color is observed, it should be cleaned and inspected and then properly relubricated.

Conveyor Drive Sprocket

Item 3 is the conveyor drive sprocket, used to drive the conveyor. The chain must engage and disengage a sprocket in at least two locations. The sprocket should be kept in good condition to prevent the chain from hanging in the sprocket and causing unnecessary wear. The sprocket and chain should be inspected to be sure engagement and disengagement is trouble-free.

Alignment is another main inspection point. If the sprockets are misaligned, wear will occur on the sides of the sprockets and inside the links of the chain. These areas need close attention.

Lubrication is also important between the sprocket and the chain. This will prevent excessive wear between the two components.

Conveyor Chain

Item 4 is the conveyor chain. Conveyor chains come in many different types, styles, and sizes. It's best to consult the manufacturer for proper maintenance procedures, but the inspections can be classed in several broad areas. The first would be lubrication. Lubrication is important to give the chain the flexibility to move around the sprocket without damaging the chain. The lubricant must be able to penetrate between all moving parts of the chain. A second consideration is the alignment of the sprockets. Misalignment causes rapid wear on the inside parts of the chain. If this type of wear occurs, the alignment should be checked.

The chain should also be examined to ensure that it isn't rubbing or dragging on any part of the supporting framework or floor coverings. This causes wear on both components and could result in enough damage to cause an operational shutdown.

Conveyor Wheels and Tracks

Items 5 and 6 are conveyor wheels and tracks. While these aren't on all conveyors, they're common in heavier industrial applications. This arrangement gives additional support to the conveyor for heavy loads. Lubrication is important to keep the wheels turning on the conveyor. If they won't turn freely, they're of no use. Lubrication is usually administered manually with a power grease gun. It's important that this be carried out on a scheduled basis to prevent wear. The inspector should be alert to the amount of lubricant evidenced on the wheels. The track should be inspected to ensure that it's straight and is supporting the wheels on the conveyor. All framework, bolts, and welds should be inspected to keep the system operating at peak efficiency. Any defect in these areas can lead to an operational shutdown.

Preventive Maintenance Inspection
Inspection 5: Typical Chain Conveyor

Check the column that indicates the condition of the unit or what problem exists named in the left column.	O.K.	Requires lubrication	Requires adjustment	Requires replacement	Requires cleaning	Excessive vibration	Excessive heat	Loose	See additional comments
1. Electric Motor									
A. Bearings									
B. Base and bolts									
C. Temperature									
D. Vibration									
E. Noise									
2. Coupling									
A. Alignment									
B. Lubrication									
3. Sprocket									
A. Teeth wear									
B. Lubrication									
C. Alignment									
D. Mounting to shaft									
4. Conveyor Chain									
A. Lubrication									
B. Wear									
5. Conveyor Wheels									
A. Lubrication									

Inspection 5 continued

Check the column that indicates the condition of the unit or what problem exists named in the left column.	O.K.	Requires lubrication	Requires adjustment	Requires replacement	Requires cleaning	Excessive vibration	Excessive heat	Loose	See additional comments
6. Track and Frame									
A. Alignment									
B. Loose bolts									
C. Broken welds									
D. Bent or broken supports									

Additional Comments:

Inspection 6: A Typical Hydraulic Lift

Refer to Figure B-6 for this section.

Pump Inlet

Item 1 is the inlet of the pump. The inlet usually has some form of filter on it. The filter should be changed on a regular basis to keep it from becoming stopped up. This will keep it in good condition so it can pass the required amount of fluid on to the pump. This inlet line should always be below the level of the fluid in the tank, thus preventing air from entering the inlet of the pump and damaging the system.

Return Line

Item 2 is the return line from the system. This line carries all the fluid returning to the reservoir. It should enter the tank and stay at a level that will not allow air to enter the returning oil. This will prevent the oil and air mixing and being carried into the system. The inspector should be alert to keep the oil at a level that will prevent splashing or churning in the tank.

Electric Motor

Item 3 is the electric motor. The most common inspection item for the motor is for heat. The motor should be approximately 25 to 29°F higher in temperature than the surrounding environment. If the motor's temperature is higher than this, it will have a shorter life due to the deteriorating effect the heat has on the insulation of the motor. Every 20°-temperature rise above the environmental temperature cuts the life of the insulation by one half. When the insulation fails, so does the motor. If the temperature rises above this level, efforts should be made to find the problem, so the motor can be cooled back down again.

Another inspection point on the motor is the bearings. The bearings support the rotating part of the motor (usually called the armature or the rotor). The inspection of the bearings should include three basic items: heat, noise, and vibration. These three indicators of trouble may appear singly or in any combination. Monitoring these three conditions may require nothing more than the use of the

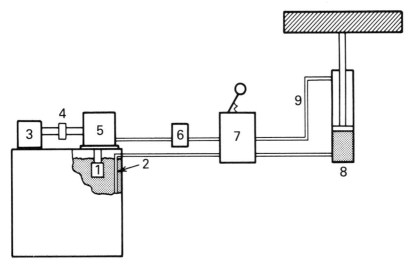

Figure B-6. For inspection 6: simple hydraulic lift.

inspector's natural senses. If the equipment is critical in the manufacturing process, it may be advisable to use some form of monitoring
or nondestructive testing equipment. For monitoring the temperature, some form of hand-held thermometer may be sufficient. If the
equipment is in a difficult-to-reach or unsafe area, some form of
temperature-monitoring device may be used. Vibration may be measured by a hand-held meter or a permanent monitoring device. Either of these types of vibration meters can determine if conditions
such as insufficient lubrication, misalignment, contamination, or normal wear are occurring in the bearing. If any of these conditions are
detected, then steps should be taken to correct the problem. If
they're not corrected, then the bearing will fail, and the replacement
will also fail quickly.

   Another item on the motor to inspect is the base. The base bolts
should always be checked to ensure that they're tightened to the
correct torque specifications. If they're not, they may come loose
during operation, allowing the motor to misalign with the remaining
parts of the drive. Periodically, the base of the motor and the mounting for the motor should be inspected. The foundation may develop
faults, allowing the motor to loosen on the base. Also, settling of the
base may occur, causing excessive stress on the motor base. The
base should always be completely clean when installing the motor.

Couplings

Item 4 is the coupling, which is used to connect two shafts together. The most critical item is the alignment of the coupling halves. If they're not aligned within very close tolerances, rapid wear will occur. A rule of thumb is that they must be within 0.005 of an inch. Rigid couplings must be in exact alignment. Flexible couplings may be able to withstand 0.005. The closer the coupling alignment is to being exact, the longer the coupling will last. Misalignment also damages related items in the drive, such as the bearings and shafts. Correct alignment of couplings cannot be overemphasized.

Lubrication is another prime consideration in coupling inspections. If correct lubrication isn't provided for flexible couplings, rapid wear and complete failure will result. If metal-to-metal contact occurs in the coupling, welding and tearing of the coupling material occurs. It's also important not to overlubricate. Overlubrication causes fluid friction, which causes heat buildup and destroys the lubricant. This then results in rapid wear of the coupling material and failure of the coupling. If a coupling is opened up for inspection and a reddish brown color is observed, it should be cleaned and inspected and then properly relubricated.

Pump

Item 5 is the pump, which is used to produce the necessary flow in the system. The inspector can usually determine the problem with the pump by listening to it. A hydraulic pump is supposed to run quietly. A growling noise indicates a problem with the system. Cavitation may be occurring or air may be getting into the inlet. A quick check of the reservoir may give an answer. If the oil looks normal and the pump is growling, it's probably cavitation. If the pump is growling and the oil is milky in the reservoir, it's probably air getting into the inlet.

The pump also has base bolts. Make sure the bolts are tight. Any looseness will allow the pump to shift and cause misalignment, which will damage the pump and motor. From time to time, alignment should be checked to ensure that no shifting has taken place. If bearing wear occurs or if wear occurs in the pumping element, the output and efficiency of the system will diminish. If this is suspect, the inspector may want to use a flow and pressure check to determine if the pump is worn enough to be replaced.

Pressure Relief Valve

Item 6 is the pressure relief valve. This valve is used to control maximum system pressure. It should be set at the level specified by the system manufacturer. This valve should be dismantled from time to time to check for worn or broken parts. If the system pressure drops or is not consistent, this valve may be the problem.

There may be an unloading valve in this area to dump the load of the pump to the reservoir at a lower pressure to prevent overheating of the pressure relief valve. If the relief valve is excessively hot, the pressure setting of the relief and unloading valve should be examined.

Directional Control Valve

Item 7 is the directional control valve. This valve can be operated by various means: electrical, pneumatic, mechanical, or manual. Its purpose is to control the direction of the actuator. One main consideration with this valve is heat. If the temperature is excessive, it indicates a possible internal leak. This will mean that wear has occurred that allows the valve to pass fluid back to the tank. It would be advisable to dismantle the valve for a visual inspection. From time to time, dirt, gums, and varnishes can build up and prevent free movement of the valve. It will be necessary to dismantle the valve to clean it.

Hydraulic Cylinder

Item 8 is the hydraulic cylinder. This is the device being controlled in the system. The cylinder should extend and retract freely with no load. If the cylinder fails to move and the rest of the system is operating correctly, it's possible for the cylinder to be passing fluid through its seals. If this is the case, it will be necessary to dismantle the cylinder and replace the seals. Other inspection points include the rod seal, which should prevent oil from flowing out of the cylinder rod, and the mounting of the cylinder. If the mounting isn't correct, it will allow too much play, causing premature wear of the seals.

Lines

Item 9 indicates the lines. The lines should always be fastened or anchored securely. This prevents movement of the lines during the operation of the system. If they're not anchored, the movement will wear the lines and cause loose fittings or broken welds. The inspector should be alert for any line leakage.

Inspection 7: A Typical Hydraulic Motor Circuit

Refer to Figure B-7 for this section.

Pump Inlet

Item 1 is the inlet of the pump. The inlet usually has some form of filter on it. The filter should be changed on a regular basis to keep it from becoming stopped up. This will keep it in good condition so it can pass the required amount of fluid on to the pump. This inlet line should always be below the level of the fluid in the tank, thus preventing air from entering the inlet of the pump and damaging the system.

Return Line

Item 2 is the return line from the system. This line carries all the fluid returning to the reservoir. It should enter the tank and stay at a level that will not allow air to enter the returning oil. This will prevent the oil and air mixing and being carried into the system. The inspector should be alert to keep the oil at a level that will prevent splashing or churning in the tank.

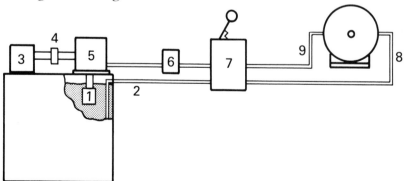

Figure B-7. For inspection 7: simple hydraulic motor.

Preventive Maintenance Inspection
Inspection 6: Hydraulic Lift

Check the column that indicates the condition of the unit or what problem exists named in the left column.	O.K.	Requires lubrication	Requires adjustment	Requires replacement	Requires cleaning	Excessive vibration	Excessive heat	Loose	See additional comments
1. Inlet Filter									
A. Clean									
B. Free intake flow									
2. Return Line									
A. Below fluid level									
3. Electric Motor									
A. Bearings									
B. Base and bolts									
C. Temperature									
D. Vibration									
E. Noise									
4. Coupling									
A. Alignment									
B. Lubrication									
5. Pump									
A. Noise									
B. Flow									
C. Pressure									
D. Base and bolts									
E. Alignment									
F. Leakage									

(continued on the following page)

Inspection 6 continued

Check the column that indicates the condition of the unit or what problem exists named in the left column.	O.K.	Requires lubrication	Requires adjustment	Requires replacement	Requires cleaning	Excessive vibration	Excessive heat	Loose	See additional comments
6. Relief Valve									
A. Adjust pressure									
B. Heat									
7. Directional Control Valve									
A. Free Operation									
B. Heat									
8. Hydraulic Cylinder									
A. Leakage									
B. Alignment									
C. Heat									
9. Lines									
A. Mounting secure									
B. Cracks									
C. Loose fittings									
Note: Depending on operating times, the hydraulic fluid should be checked and analyzed to prevent wear on the system.									

Additional Comments:

Electric Motor

Item 3 is the electric motor. The most common inspection item for the motor is for heat. The motor should be approximately 25 to 29°F higher in temperature than the surrounding environment. If the motor's temperature is higher than this, it will have a shorter life due to the deteriorating effect the heat has on the insulation of the motor. Every 20°-temperature rise above the environmental temperature cuts the life of the insulation by one half. When the insulation fails, so does the motor. If the temperature rises above this level, efforts should be made to find the problem, so the motor can be cooled back down again.

Another inspection point on the motor is the bearings. The bearings support the rotating part of the motor (usually called the armature or the rotor). The inspection of the bearings should include three basic items: heat, noise, and vibration. These three indicators of trouble may appear singly or in any combination. Monitoring these three conditions may require nothing more than the use of the inspector's natural senses. If the equipment is critical in the manufacturing process, it may be advisable to use some form of monitoring or nondestructive testing equipment. For monitoring the temperature, some form of hand-held thermometer may be sufficient. If the temperature is in a difficult-to-reach or unsafe area, some form of temperature-monitoring device may be used. Vibration may be measured by a hand-held meter or a permanent monitoring device. Either of these types of vibration meters will be able to determine if conditions such as insufficient lubrication, misalignment, contamination, or normal wear are occurring in the bearing. If any of these conditions are detected, then steps should be taken to correct the problem. If they're not corrected, then the bearing will fail, and the replacement will also fail quickly.

Another item on the motor to inspect is the base. The base bolts should always be checked to ensure that they're tightened to the correct torque specifications. If they're not, they may come loose during operation, allowing the motor to misalign with the remaining parts of the drive. Periodically, the base of the motor and the mounting for the motor should be inspected. It's possible that the foundation may develop faults, allowing the motor to loosen on the base. Also, settling of the base may occur, causing excessive stress on the motor base. The base should always be completely clean when installing the motor.

Couplings

Item 4 is the coupling, which is used to connect two shafts together. The most critical item is the alignment of the coupling halves. If they're not aligned within very close tolerances, rapid wear will occur. A rule of thumb is that they must be within 0.005 of an inch. Rigid couplings must be in exact alignment. Flexible couplings may be able to withstand 0.005. The closer the coupling alignment is to being exact, the longer the coupling will last. Misalignment also damages related items in the drive, such as the bearings and shafts. Correct alignment of couplings cannot be overemphasized.

Lubrication is another prime consideration in coupling inspections. If correct lubrication isn't provided for flexible couplings, rapid wear and complete failure will result. If metal-to-metal contact occurs in the coupling, welding and tearing of the coupling material occurs. It's also important not to overlubricate. Overlubrication causes fluid friction, which builds up heat and destroys the lubricant, resulting in rapid wear of the coupling material and failure of the coupling. If a coupling is opened up for inspection and a reddish brown color is observed, it should be cleaned and inspected and then properly relubricated.

Pump

Item 5 is the pump. The pump is used to produce the necessary flow in the system. The inspector can usually determine the problem with the pump by listening to it. A hydraulic pump is supposed to run quietly. A growling noise indicates a problem with the system. Cavitation may be occurring or air may be getting into the inlet. A quick check of the reservoir may give an answer. If the oil looks normal and the pump is growling, it's probably cavitation. If the pump is growling and the oil is milky in the reservoir, it's probably air getting into the inlet. The pump also has base bolts, which should be tight. Any looseness will allow the pump to shift and cause misalignment, which damages the pump and motor. From time to time, alignment should be checked to ensure that no shifting has taken place. If bearing wear occurs or if wear occurs in the pumping element, the output and efficiency of the system will diminish. If this is suspect, the inspector may want to use a flow and pressure check to determine if the pump is worn enough to be replaced.

Pressure Relief Valve

Item 6 is the pressure relief valve, which is used to control maximum system pressure. It should be set at the level specified by the system manufacturer. This valve should be dismantled from time to time to check for worn or broken parts. If the system pressure drops or isn't consistent, this valve may be the problem.

Also, there may be an unloading valve in this area to dump the load of the pump to the reservoir at a lower pressure to prevent overheating of the pressure relief valve. If the relief valve is excessively hot, the pressure setting of the relief and unloading valve should be examined.

Directional Control Valve

Item 7 is the directional control valve. This valve can be operated by various means: electrical, pneumatic, mechanical, or manual. Its purpose is to control the direction of the actuator. One main consideration with this valve is heat. Excessive temperature indicates a possible internal leak. Such a condition means that wear has occurred which allows the valve to pass fluid back to the tank. It's advisable to dismantle the valve for a visual inspection. Dirt, gums, and varnishes build up from time to time and prevent free movement of the valve. In this case, it's necessary to dismantle the valve to clean it.

Hydraulic Motor

Item 8 in Figure B-7 is the hydraulic motor. This is the device being controlled in the system. The motor should rotate freely with no load. If the motor fails to move and the rest of the system is operating correctly, it's possible that the motor is passing fluid through its seals. If this is the case, it will be necessary to dismantle the motor and replace the seals. Another possibility is that the internal components have worn to a degree that allows free passage of fluid. If this is the case, replacement of the components or the motor will be required. Other inspection points include the shaft seal, which should prevent oil from flowing out of the motor shaft, and the mounting of the motor. Incorrect mounting allows too much play and causes premature wear of the seals.

Lines

Item 9 denotes the lines. The lines should always be fastened or anchored securely to prevent their movement during operation of the system. If they're not anchored, the movement will wear the lines and cause loose fittings or broken welds. The inspector should be alert for any line leakage.

Inspection 8: Typical Pneumatic Cylinder

Refer to Figure B-8 for this section.

Electric Motor

Item 1 in the figure is the electric motor. The most common inspection item for the motor is for heat. The motor should be approximately 25 to 29°F higher in temperature than the surrounding environment. If the motor's temperature is higher than this, it will have a shorter life due to the deteriorating effect the heat has on the insulation of the motor. Every 20°-temperature rise above the environmental temperature cuts the life of the insulation by one half. When the insulation fails, so does the motor. If the temperature rises above this level, efforts should be made to find the problem, so the motor can be cooled back down again.

Another inspection point on the motor is the bearings, which support the rotating part of the motor (usually called the armature or

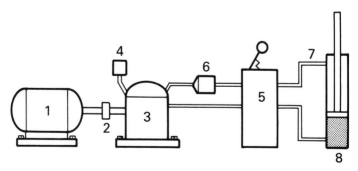

Figure B-8. For inspection 8: simple pneumatic cylinder.

Preventive Maintenance Inspection
Inspection 7: Hydraulic Motor

Check the column that indicates the condition of the unit or what problem exists named in the left column.	O.K.	Requires lubrication	Requires adjustment	Requires replacement	Requires cleaning	Excessive vibration	Excessive heat	Loose	See additional comments
1. Inlet Filter									
A. Clean									
B. Free intake flow									
2. Return Line									
A. Below fluid level									
3. Electric Motor									
A. Bearings									
B. Base and bolts									
C. Temperature									
D. Vibration									
E. Noise									
4. Coupling									
A. Alignment									
B. Lubrication									
5. Pump									
A. Noise									
B. Flow									
C. Pressure									
D. Base and bolts									
E. Alignment									
F. Leakage									

Inspection 7 continued

Check the column that indicates the condition of the unit or what problem exists named in the left column.	O.K.	Requires lubrication	Requires adjustment	Requires replacement	Requires cleaning	Excessive vibration	Excessive heat	Loose	See additional comments
6. Relief Valve									
A. Adjust pressure									
B. Heat									
7. Directional Control Valve									
A. Free Operation									
B. Heat									
8. Hydraulic Motor									
A. Leakage									
B. Alignment									
C. Heat									
9. Lines									
A. Mounting secure									
B. Cracks									
C. Loose fittings									
Note: Depending on operating times, the hydraulic fluid should be checked and analyzed to prevent wear on the system.									

Additional Comments:

the rotor). The inspection of the bearings should include three basic items: heat, noise, and vibration. These three indicators of trouble may appear singly or in any combination. Monitoring these three conditions may require nothing more than the use of the inspector's natural senses. If the equipment is critical in the manufacturing process, it may be advisable to use some form of monitoring or nondestructive testing equipment. For monitoring the temperature, some form of hand-held thermometer may be sufficient. If the equipment is in a difficult-to-reach or unsafe area, some form of temperature-monitoring device may be used. Vibration may be measured by a hand-held meter or a permanent monitoring device. Either of these types of vibration meters can determine if conditions such as insufficient lubrication, misalignment, contamination, or normal wear are occurring in the bearing. If any of these conditions are detected, then steps should be taken to correct the problem. If they're not corrected, the bearing will fail, and the replacement will also fail quickly.

Another item on the motor to inspect is the base. The base bolts should always be checked to ensure that they're tightened to the correct torque specifications. If they're not, they may come loose during operation, allowing the motor to misalign with the remaining parts of the drive. Periodically, the base of the motor and the mounting for the motor should be inspected. The foundation may develop faults, allowing the motor to loosen on the base; or settling of the base may occur, causing excessive stress on the motor base. The base should always be completely clean when installing the motor.

Couplings

Item 2 is the coupling, which is used to connect two shafts together. The most critical item is the alignment of the coupling halves. If they're not aligned within very close tolerances, rapid wear will occur. A rule of thumb is that they must be within 0.005 of an inch. Rigid couplings must be in exact alignment. Flexible couplings may be able to withstand 0.005. The closer the coupling alignment is to being exact, the longer the coupling will last. Misalignment also damages related items in the drive, such as the bearings and shafts. Correct alignment of couplings cannot be overemphasized.

Lubrication is another prime consideration in coupling inspections. If correct lubrication isn't provided for flexible couplings, rapid wear and complete failure will result. If metal-to-metal contact occurs in the coupling, welding and tearing of the coupling material occurs. It's also important not to overlubricate. Overlubrication causes fluid friction, which builds up heat and destroys the lubricant. In this case, rapid wear of the coupling material and failure of the coupling will occur. If a coupling is opened up for inspection and a reddish brown color is observed, it should be cleaned and inspected and then properly relubricated.

Compressor

Item 3 is the compressor, which changes the mechanical energy into pneumatic energy. Most compressors have a pumping chamber equipped with valves and seals to properly compress the air. If either of the valves (inlet or discharge) or the seals are leaking, the output of the pump won't be as high as it should be. If the compressor cannot keep up with system demand, but is rated high enough, the inspector should recommend dismantling and rebuilding the compressor. Most compressors have a crankcase or gearcase for the storage of lubricant. The lubricant should always be kept at the proper level to prevent damage to the mechanical components in the compressor.

Some compressors are equipped with internal coolers to lower the temperature of the air. Coolers may be air or water cooled. If excessive moisture is in the air after compression, the cooler should be checked for leakage. If the air is still hot after being run through the cooler, the inspector should check the cooler to see if it's stopped up and not allowing good flow of the cooling medium.

Inlet Filter

Item 4 is the inlet filter. All air in the system must at one time come through this filter. The inlet should always be positioned in a clean, dry location for efficient operation of the system. The filter should be periodically changed, otherwise the compressor can't draw enough air through the filter, and the system will operate at a high temperature and in a sluggish manner. Any contamination that passes through this filter is sure to cause problems in the system.

Directional Control Valve

Item 5 is the directional control valve. This component controls the direction of the actuator and may be activated by mechanical, pneumatic, or electrical means. The valve has a spool inside that shifts, directing the air flow to an outlet port. The port out of which the air flow is directed determines the direction of the device being controlled. If the directional control valve doesn't shift freely, it should be dismantled and inspected. The valve should be cleaned and all defective parts replaced before it is reassembled.

Muffler

Item 6 is the muffler. It's used to quiet the exhaust of the system as it's returned to the atmosphere. There are usually two main types of mufflers. One is freeflow and the other is adjustable. The adjustable one can be used to control the flow rate of the air as it's returned to the atmosphere. This will allow for speed control of the actuator in the circuit. If either type becomes clogged or restricted, it will affect operation of the circuit. The muffler should always be inspected for proper operation.

Lines Carrying Compressed Air

Item 7 shows the lines carrying the compressed air. The lines and related fittings should always be checked for leakage. If the lines are allowed to leak, the compressor won't be able to supply enough air to operate the system properly. There are some listening devices on the market that will detect air leaks at considerable distances. If the lines are in locations that are difficult to reach, these devices are very convenient. The lines should also be checked for rigidity. They should be clamped and held in place to prevent motion and vibration, both of which will cause eventual loosening or cracking of the lines.

Pneumatic Cylinder

Item 8 is the pneumatic cylinder. Pneumatic cylinders are usually used for high-speed low-power applications. Inspection points include proper motion, looseness of mounting devices, leakage of any fittings, and leakage at the rod of the cylinder. Any of these defects need attention as soon as it's possible to schedule the repair.

Preventive Maintenance Inspection
Inspection 8: Pneumatic Cylinder

Check the column that indicates the condition of the unit or what problem exists named in the left column.	O.K.	Requires lubrication	Requires adjustment	Requires replacement	Requires cleaning	Excessive vibration	Excessive heat	Loose	See additional comments
1. Electric Motor									
A. Bearings									
B. Base and bolts									
C. Temperature									
D. Vibration									
E. Noise									
2. Coupling									
A. Alignment									
B. Lubrication									
3. Compressor									
A. Flow									
B. Pressure									
C. Noise									
D. Vibration									
E. Lubrication									
F. Heat									
4. Inlet Filter									
A. Clean									
B. Free flow									
5. Directional Control Valve									
A. Free movement									
B. Proper air flow									

Inspection 8 continued

Check the column that indicates the condition of the unit or what problem exists named in the left column.	O.K.	Requires lubrication	Requires adjustment	Requires replacement	Requires cleaning	Excessive vibration	Excessive heat	Loose	See additional comments
6. Muffler									
A. Quiet									
B. Good flow									
7. Lines									
A. Properly mounted									
B. Leaks									
C. Loose fittings									
D. Broken piping									
8. Pneumatic Cylinder									
A. Free movement									
B. Good alignment									
C. Proper mounting									
D. Leakage									

Additional Comments:

Inspection 9: Typical Pneumatic Motor

Refer to Figure B-9 for this section.

Electric Motor

Item 1 is the electric motor. The most common inspection item for the motor is for heat. The motor should be approximately 25 to 29°F higher in temperature than the surrounding environment. If the motor's temperature is higher than this, it will have a shorter life due to the deteriorating effect the heat has on the insulation of the motor. Every 20°-temperature rise above the environmental temperature cuts the life of the insulation by one half. When the insulation fails, so does the motor. If the temperature rises above this level, efforts should be made to find the problem, so the motor can be cooled back down again.

Another inspection point on the motor is the bearings. The bearings support the rotating part of the motor (usually called the armature or the rotor). The inspection of the bearings should include three basic items: heat, noise, and vibration. These three indicators of trouble may appear singly or in any combination. Monitoring these three conditions may require nothing more than the use of the inspector's natural senses. If the equipment is critical in the manufacturing process, it may be advisable to use some form of monitoring or nondestructive testing equipment. For monitoring the temperature, some form of hand-held thermometer may be sufficient. If the equipment is in a difficult-to-reach or unsafe area, some form of temperature-monitoring device may be used. Vibration may be measured by a hand-held meter or a permanent monitoring device. Either of these types of vibration meters can determine if conditions such as insufficient lubrication, misalignment, contamination, or nor-

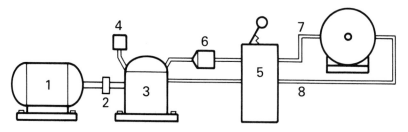

Figure B-9. For inspection 9: simple pneumatic motor.

mal wear are occurring in the bearing. If any of these conditions are detected, then steps should be taken to correct the problem. If they're not corrected, the bearing will fail, and the replacement will also fail quickly.

Another item on the motor to inspect is the base. The base bolts should always be checked to ensure that they're tightened to the correct torque specifications. If they're not, they may come loose during operation, allowing the motor to misalign with the remaining parts of the drive. Periodically, the base of the motor and the mounting for the motor should be inspected. The foundation may develop faults, allowing the motor to loosen on the base. Also, settling of the base may occur, causing excessive stress on the motor base. The base should always be completely clean when installing the motor.

Couplings

Item 2 is the coupling, which is used to connect two shafts together. The most critical item is the alignment of the coupling halves. If they're not aligned within very close tolerances, rapid wear will occur. A rule of thumb is that they must be within 0.005 of an inch. Rigid couplings must be in exact alignment. Flexible couplings may be able to withstand 0.005. The closer the coupling alignment is to being exact, the longer the coupling will last. Misalignment, however, can damage related items in the drive, such as the bearings and shafts. Correct alignment of couplings cannot be overemphasized.

Lubrication is another prime consideration in coupling inspections. If correct lubrication isn't provided for flexible couplings, rapid wear and complete failure will result. If metal-to-metal contact occurs in the coupling, welding and tearing of the coupling material occurs. Along this same line, it's also important not to overlubricate. Overlubrication causes fluid friction, which builds up heat and destroys the lubricant. This condition results in rapid wear of the coupling material and failure of the coupling. If a coupling is opened up for inspection and a reddish brown color is observed, it should be cleaned and inspected and then properly relubricated.

Compressor

Item 3 is the compressor, the device that changes the mechanical energy into pneumatic energy. Most compressors have a pumping chamber equipped with valves and seals to properly compress the air. If either of the valves (inlet or discharge) or the seals are leaking, the output of the pump won't be as high as it should. If the compressor cannot keep up with system demand, but is rated high enough, the inspector should recommend dismantling and rebuilding the compressor. Most compressors have a crankcase or gearcase for the storage of lubricant. The lubricant should always be kept at the proper level to prevent damage to the mechanical components in the compressor. Some compressors are equipped with internal coolers to lower the temperature of the air. Coolers may be air or water cooled. If excessive moisture is in the air after compression, the cooler should be checked for leakage. If the air is still hot after being run through the cooler, the inspector should check the cooler to see if it's stopped up and not allowing good flow of the cooling medium.

Inlet Filter

Item 4 is the inlet filter. This filter is important because all air in the system must at one time come through this filter. The inlet should always be positioned in a clean, dry location for efficient operation of the system. The filter should be periodically changed, otherwise the compressor can't draw enough air through the filter, and the system will operate at a high temperature and in a sluggish manner. Any contamination that passes through this filter is sure to cause problems in the system.

Directional Control Valve

Item 5 is the directional control valve. This component controls the direction of the actuator and may be activated by mechanical, pneumatic, or electrical means. The valve has a spool inside that shifts, directing the air flow to an outlet port. The port out of which the air flow is directed determines the direction of the device being controlled. If the directional control valve doesn't shift freely, it should be dismantled and inspected. The valve should be cleaned and all defective parts replaced before it is reassembled.

Muffler

Item 6 is the muffler, which is used to quiet the exhaust of the system as it's returned to the atmosphere. There are usually two main types of mufflers. One is free flow and the other is adjustable. The adjustable one can be used to control the flow rate of the air as it's returned to the atmosphere. This will allow for speed control of the actuator in the circuit. If either type becomes clogged or restricted, it will affect operation of the circuit. The muffler should always be inspected for proper operations.

Lines Carrying Compressed Air

Item 7 shows the lines carrying the compressed air. The lines and related fittings should always be checked for leakage. If the lines are allowed to leak, the compressor won't be able to supply enough air to operate the system properly. There are some listening devices on the market that detect air leaks at considerable distances. If the lines are in locations that are difficult to reach, these devices are very convenient. The lines should also be checked for rigidity. They should be clamped and held in place to prevent motion and vibration, both of which will cause eventual loosening or cracking of the lines.

Pneumatic Motor

Item 8 is the pneumatic motor. Pneumatic motors are usually used for high-speed low-torque applications. Inspection points include proper motion, looseness of mounting devices, leakage of any fittings, and leakage at the rod of the cylinder. Any of these defects need attention as soon as it's possible to schedule the repair.

Inspection 10: A Simple Electrical Starter

Refer to Figure B-10 for this section.

Incoming Wiring

Item 1 is the incoming wiring. This wiring should be inspected for worn or defective insulation. Any points where the wires make sharp bends or touch any metal supports are good places to begin inspecting. Frayed insulation or exposed wiring is a condition needing immediate attention. The wires should also be inspected where they fasten to the starter, whether this is on a terminal block, or if they fasten directly to the relay. The wires should be checked for

Preventive Maintenance Inspection
Inspection 9: Pneumatic Motor

Check the column that indicates the condition of the unit or what problem exists named in the left column.	O.K.	Requires lubrication	Requires adjustment	Requires replacement	Requires cleaning	Excessive vibration	Excessive heat	Loose	See additional comments
1. Electric Motor									
A. Bearings									
B. Base and bolts									
C. Temperature									
D. Vibration									
E. Noise									
2. Coupling									
A. Alignment									
B. Lubrication									
3. Compressor									
A. Flow									
B. Pressure									
C. Noise									
D. Vibration									
E. Lubrication									
F. Heat									
4. Inlet Filter									
A. Clean									
B. Free flow									
5. Directional Control Valve									
A. Free movement									
B. Proper air flow									

Inspection 9 continued

Check the column that indicates the condition of the unit or what problem exists named in the left column.	O.K.	Requires lubrication	Requires adjustment	Requires replacement	Requires cleaning	Excessive vibration	Excessive heat	Loose	See additional comments
6. Muffler									
A. Quiet									
B. Good flow									
7. Lines									
A. Properly mounted									
B. Leaks									
C. Loose fittings									
D. Broken piping									
8. Pneumatic Motor									
A. Free movement									
B. Good alignment									
C. Proper mounting									
D. Leakage									

Additional Comments:

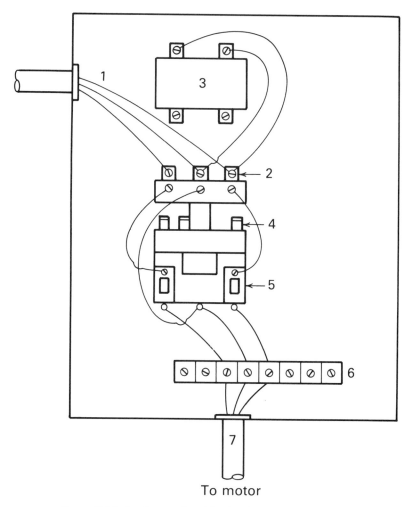

Figure B-10. For inspection 10: simple electrical starter.

looseness, because this causes heating and discoloration of the terminal. Discoloration means that the fitting is loose or has been loose, and thus should be closely checked for proper tightness.

Terminals in the Starter Panel

Item 2 represents the terminals in the starter panel. The main inspection point of the terminals is tightness. If they work loose, then high current develops at this point. High current develops high heat, which may be intense enough to weld the terminal. If this occurs, replacement of the terminal is required. Any terminal that has been overheated will be discolored, which is a good visible sign to the inspector.

Control Transformer

Item 3 is the control transformer, a device used to lower the line voltage. The two most common industrial sizes are 220 to 110 V and 440 to 220 V. The terminals and the insulation of the transformer are the two main inspection points. Correct tightness of the terminals is important. The insulation on the transformer should be visually inspected to ensure that it's not breaking or peeling off the transformer. This condition exposes the wiring and presents a possible hazard or malfunction. All insulation should be kept in good condition.

Contact Tips

Item 4 shows the contact tips in the relay. In some cases, it's necessary to remove a shield to inspect these tips, which are usually made of a silver-coated copper alloy. If they're worn sufficiently, replacement is necessary to ensure proper operation of the relay. Springs are used to compensate for wear of the tips. These should be in good condition and in their proper location, otherwise the relay won't operate correctly, and burning of the tips will result. If the tips are sticking or welding together, the spring tension should be checked. If the tips stick together periodically, replacement should be considered.

Overload Coil and Tips

Item 5 is the overload coil and tips. The line current is passed through these tips. If the current becomes excessive, the heat will open these tips and drop the relay out, preventing overloads. The problem occurs when the overload is changed. If too large an overload is put in, it will damage the system and never drop the relay out. If too small an overload is installed, it will open the circuit too often, causing more service calls than necessary. The inspector should check to be sure that the correct-sized overload is in the relay. Tightness of all connections is a must if the current through the overload is to be a proper indication of current in the circuit.

Terminal Strip

Item 6 is the terminal strip, a device that connects the relay panel to the wiring in the rest of the circuit. Tightness of the connections on the strip is very important. Also, the wires should be spaced far enough apart so that as they go to different terminals, they don't make contact. Neat, well-arranged wiring is very helpful to the inspector and repairman.

Wiring and Related Hardware

Item 7 is the wiring and related hardware going to a motor. The wiring should be in good condition, with good insulation. The conduit or flexible covering for the wires should be firmly anchored to prevent unnecessary motion or vibration. This will prevent additional wear on the wires. All devices should be securely mounted in the panel. If the components are allowed to move, they may short out against each other or flex the wire enough to break it.

Inspection 11: A Typical Lighting Circuit

Refer to Figure B-11 for this section.

Circuit Breaker

Item 1 is the circuit breaker that supplies power to the circuit. The main inspection point of this component is the tightness of all connections. Tight connections prevent most problems. If the breaker begins tripping frequently, it's usually a sign of wear in the breaker. If there is no problem in the circuit, consideration may be given to rebuilding or replacing the breaker.

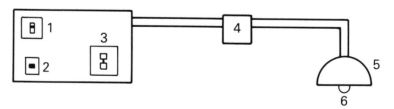

Figure B-11. For inspection 11: typical lighting circuit.

Preventive Maintenance Inspection
Inspection 10: Electrical Starter

Check the column that indicates the condition of the unit or what problem exists named in the left column.	O.K.	Requires lubrication	Requires adjustment	Requires replacement	Requires cleaning	Excessive vibration	Excessive heat	Loose	See additional comments
1. Incoming Wiring									
A. Insulation									
B. Loose connections									
C. Heat									
2. Terminals									
A. Proper mountings									
B. Loose connections									
C. Heat									
3. Transformer									
A. Proper mounting									
B. Insulation									
C. Loose connections									
D. Heat									
4. Relay Tips									
A. Wear									
B. Springs									
C. Heat									
5. Overload Relay									
A. Proper size									
B. Tip condition									

Inspection 10 continued

Check the column that indicates the condition of the unit or what problem exists named in the left column.	O.K.	Requires lubrication	Requires adjustment	Requires replacement	Requires cleaning	Excessive vibration	Excessive heat	Loose	See additional comments
C. Loose connections									
D. Heat									
6. Terminal Strip									
A. Correct mounting									
B. Loose connections									
C. Heat									
7. Outside Leads									
A. Insulation									
B. Loose connections									
C. Heat									

Additional Comments:

On–Off Switch

Item 2 is the on–off switch for the lights. The main inspection point here is the correct connection and protection of the wiring. If the wires are connected and tightened correctly, most problems will be minimal. The wiring should be in conduit or some form of flexible covering to prevent damage to the wires. If they're not protected, any small breakdown in insulation will cause a lighting failure.

Relay Panel

Item 3 is the relay panel. This item is optional, although it is found in heavier industrial lighting. The relay panel should be given the same basic inspection as the relay panel in the previous example. Almost all components will be similar.

Ballast

Item 4 is a ballast of the transformer. This device is used to change the line voltage to the correct level for the light. Proper connection of the wiring and tightness of all fittings is the primary inspection.

Light Shade and Socket

Item 5 represents the light shade and socket. The shade may require cleaning from time to time, especially in dirty environments. The socket requires occasional inspection (usually during the lamp replacement). Loose wires, screws, or connecting components are obvious signs of trouble. The mounting should always be adequate, for if the light vibrates, the filament in the bulb won't last very long.

Light Bulb

Item 6 is the light bulb. It should always be the correct wattage and voltage for the light. A larger wattage can cause an overload on the socket or too large a load on the rest of the system. A light that's too small could result in insufficient lighting and unsafe conditions.

Preventive Maintenance Inspection
Inspection 11: Typical Lighting Circuit

Check the column that indicates the condition of the unit or what problem exists named in the left column.	O.K.	Requires lubrication	Requires adjustment	Requires replacement	Requires cleaning	Excessive vibration	Excessive heat	Loose	See additional comments
1. Circuit Breaker									
A. Wire insulation									
B. Loose terminals									
C. Heat									
2. Off-On Switch									
A. Wire insulation									
B. Loose terminals									
C. Heat									
3. Relay Panel									
A. See Previous Inspection (#10)									
4. Ballast									
A. Wire insulation									
B. Loose terminals									
C. Heat									
5. Light Shade and Socket									
A. Cleanliness									
B. Loose connections									
C. Insulation									
D. Heat									

Inspection 11 continued

Check the column that indicates the condition of the unit or what problem exists named in the left column.	O.K.	Requires lubrication	Requires adjustment	Requires replacement	Requires cleaning	Excessive vibration	Excessive heat	Loose	See additional comments
6. Light Bulb									
A. Loose connections									
B. Insulation									
C. Correct wattage									

Additional Comments:

Inspection 12: Typical Motor–Generator Set

Refer to Figure B-12 for this section.

Electric Motor

Item 1 is the electric motor. The most common inspection item for the motor is for heat. The motor should be approximately 25 to 29°F higher in temperature than the surrounding environment. If the motor's temperature is higher than this, it will have a shorter life due to the deteriorating effect the heat has on the insulation of the motor. Every 20°-temperature rise above the environmental temperature cuts the life of the insulation by one half. When the insulation fails, so does the motor. If the temperature rises above this level, efforts should be made to find the problem, so the motor can be cooled back down again.

Another inspection point on the motor is the bearings, which support the rotating part of the motor (usually called the armature or the rotor). The inspection of the bearings should include three basic items: heat, noise, and vibration. These three indicators of trouble may appear singly or in any combination. Monitoring these three conditions may require nothing more than the use of the inspector's natural senses. If the equipment is critical in the manufacturing pro-cess, it may be advisable to use some form of monitoring or nonde-structive testing equipment. For monitoring the temperature, some form of hand-held thermometer may be sufficient. If the equipment is in an area that's difficult to reach or is unsafe, some form of tem-perature-monitoring device may be used. Vibration may be mea-sured by a hand-held meter or a permanent monitoring device. Ei-ther of these types of vibration meters is able to determine if conditions such as insufficient lubrication, misalignment, contamina-tion, or normal wear are occurring in the bearing. If any of these

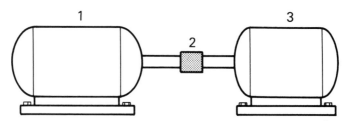

Figure B-12. For inspection 12: motor–generator set.

conditions are detected, then steps should be taken to correct the problem. If they're not corrected, then the bearing will fail, and the replacement will also fail quickly.

Another item on the motor to inspect is the base. The base bolts should always be checked to ensure that they're tightened to the correct torque specifications. If they're not, they may come loose during operation, allowing the motor to misalign with the remaining parts of the drive. Periodically, the base of the motor and the mounting for the motor should be inspected. The foundation may develop faults, allowing the motor to loosen on the base. Settling of the base may occur, causing excessive stress on the motor base. The base should always be completely clean when installing the motor.

Couplings

Item 2 is the coupling, which is used to connect two shafts together. The most critical inspection item is the alignment of the coupling halves. If they're not aligned within very close tolerances, rapid wear will occur. A rule of thumb is that they must be within 0.005 of an inch. Rigid couplings must be in exact alignment. Flexible couplings may be able to withstand 0.005. The closer the coupling alignment is to being exact, the longer the coupling will last. Misalignment will also damage related items in the drive, such as the bearings and shafts. Correct alignment of couplings cannot be overemphasized.

Lubrication is another prime consideration in coupling inspections. If correct lubrication isn't provided for flexible couplings, rapid wear and complete failure will result. If metal-to-metal contact occurs in the coupling, welding and tearing of the coupling material occurs. It's important not to overlubricate. Overlubrication causes fluid friction, which creates heat buildup and destroys the lubricant. This condition results in rapid wear of the coupling material and failure of the coupling. If a coupling is opened up for inspection and a reddish brown color is observed, it should be cleaned and inspected and then properly relubricated.

Generator

Item 3 is the generator, which is the output of the system. The usual inspection points include the armature, the rotor, and the brushes. In addition to these areas, the motor inspection points should be considered. A generator is usually nothing more than a motor that is mechanically driven.

Preventive Maintenance Inspection
Inspection 12: Typical Motor-Generator Set

Check the column that indicates the condition of the unit or what problem exists named in the left column.	O.K.	Requires lubrication	Requires adjustment	Requires replacement	Requires cleaning	Excessive vibration	Excessive heat	Loose	See additional comments
1. Electric Motor									
A. Bearings									
B. Base and bolts									
C. Temperature									
D. Vibration									
E. Noise									
2. Coupling									
A. Alignment									
B. Lubrication									
3. Generator									
A. All of electric motor (#1)									
B. Armature									
C. Brushes									
D. Rotor									

Additional Comments:

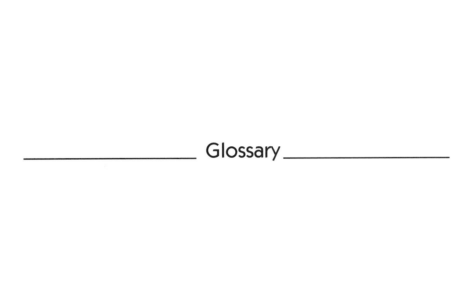

Glossary

Glossary

Abrasive — Causing a wearing down or grinding away by friction.

Absolute — A pressure scale that begins at 0 psi and at which atmospheric pressure is 14.7 psi.

Acceleration — The rate of change of an object's velocity.

Addendum — The distance from the pitch line to the tip of the tooth in a gear.

Additive — A substance added to a lubricant to improve or enhance a quality.

Alignment — The proper positioning of two components to one another.

Backlash — The amount of clearance between two gear teeth in mesh.

Bevel — A gear with teeth that are cut into the face of the gear. The gears may intersect at almost any angle.

Bleeding — The process of an oil working out of the base in a grease.

Bolt — A fastening device used with a nut to hold two or more parts together. The bolt is tightened by turning the nut.

Bushing — A cylindrical device used to reduce the friction between two moving parts.

Cavitation — A process where dissolved air is removed from the fluid on the inlet side of the pump and implodes back into the fluid on the outlet side of the pump.

Chordal thickness — The thickness of a gear tooth measured on the pitch line. A straight line measurement.

Circular thickness — The thickness of a gear tooth measured on the pitch circle. This measurement is an arc. Usually only given on spur gears.

Clearance — The radial distance between the tip of a gear tooth and the bottom of the mating tooth space.

Compression packing — Packing that accomplishes sealing by being deformed under pressure.

Corrosion — A process in which material is worn away gradually, usually by some form of a chemical action.

Coupling — A device used to connect shafting.

Cylinder — A fluid power component used for linear motion. Sometimes called a linear actuator.

Deceleration — The process of reducing an object's velocity.

Dedendum — The distance from the pitch circle to the root of a tooth.

Density — The mass of a material for a given volume.

Depth of engagement — The radial engagement of a screw thread.

Dropping point — The temperature at which a grease liquifies.

Dynamic — A type of friction relating to objects in motion.

Efficiency — A ratio of the input energy to the output energy. Usually expressed as a percentage.

End-play — Motion along the axis of a shaft.

Fit — A designation used to indicate the closeness of two mating screw threads.

Flash point — The temperature at which a substance will burst into flames.

Friction — The resistance to motion of two bodies in contact. The three types are static, dynamic, and rolling.

Gauge — A pressure scale. This scale ignores atmospheric pressure, which is 0 psig.

Gear — A mechanical toothed wheel that provides a drive with a positive transmission of torque.

Helical — A form of a gear tooth that is cut at a helix angle on the face of the gear.

Herringbone — A gear with two sets of teeth with opposite hands cut into the face. The helix angles of both sets of teeth are the same.

Hub — The center part of a coupling or sprocket.

Humidity — The measure of the amount of water vapor in a given volume of air.

Hydrodynamic — A physical property by which a rotating body can develop pressure in a surrounding fluid. Some bearings depend on this principle to support their load.

Hydrostatic — A type of bearing that has fluid pressure supplied to it by an external source.

Hypoid bevel — A bevel gear that has offset axis. The shafting may be extended to provide more support for the gear.

Input — Power or speed that is put into some mechanical or fluid power component.

Interference — A form of wear in a gear drive where two gears are in tooth tip-to-root contact.

Internal gear — A gear with the teeth cut on the inside circumference of the pitch circle.

Keyway — A groove or channel cut into a mechanical component for a key.

Lantern ring — A ring in a stuffing box used to provide lubrication for the packing.

Lead — The amount of axial distance traveled by the turning of the threaded component in one turn.

Length of engagement — The axial length that two threaded components are in contact.

Lubricant — A substance that is introduced between two or more moving components to reduce friction and wear.

Major diameter — The outside diameter of a screw thread, measured radially.

Mass — The measure of the amount of material in an object.

Mesh — The size of one of the openings in a filter or strainer; also, the working contact of gear teeth.

Micron — A unit of measure equal to 0.00039 inch.

Minor diameter — The smallest diameter of a screw thread measured radially at the root of the thread.

Motor — A hydraulic device for converting flow to rotary motion. May also be a prime mover in a mechanical or fluid power system (electric motor).

Multiple threads — A threaded device having more than one set of threads progressing along its axial length.

Nominal — A term used in rating a filter that is its approximate size in microns.

Nonpositive displacement — A condition in which a pump or compressor does not produce a given volume per revolution.

Number of threads — The number of threads per inch of axial length on a screw thread.

Orifice — An opening that restricts flow in a fluid power system.

Output — The power or speed delivered by a mechanical or fluid power drive.

Oxidation — The process of material combining with oxygen. Usually results in the formation of rust.

Penetration — A test to check the thickness of grease by dropping a fixed weight (cone shaped) into the grease from a given height. The depth the cone penetrates is the grease's penetration rating.

Pin — A type of chain link used to connect two roller links.

Pinion — The smaller of two gears in a gear drive.

Pitch — The distance from a point on one gear tooth to the corresponding point on the next tooth. In a chain drive, a distance from a point on one chain link to the corresponding point on the next link.

Pitch diameter — The diameter of an imaginary circle that connects all the pitch points on a gear or sprocket.

Planetary — A type of internal gear drive having several internal gears rotating around a center (sun) gear.

Pneumatics — A form of power transmission that uses a gas for the transmitting medium.

Positive displacement — A type of pump or compressor that displaces a certain volume for every revolution.

Power — A force moving through a distance in a given time.

Predictive maintenance — A form of maintenance that predicts the wear and need for repair before a breakdown occurs.

Pressure — Force per unit of area.

Preventive maintenance — The philosophy of maintenance that works to prevent breakdowns before they occur.

Pumpability — The ability of a lubricant to be pumped.

Rack and pinion — A type of gear drive that translates rotary motion to linear motion, or vice versa.

Radial — A type of load applied 90 degrees to the shaft axis.

Relative humidity — The ratio of the amount of water vapor that a given volume of air contains compared to the amount of water vapor that it could contain. Expressed in a percentage.

Roller — A type of chain link that is composed of a bushing, link plate, and a roller.

Root — In a gear drive, it is the bottom area between the two teeth. In a screw thread, it is the lowest point between two threads.

Runout — Movement along the axis of an object, usually a shaft.

Sprocket — A wheel that has teeth cut in its outside circumference for engagement with a chain.

Thrust — A type of load that is applied along a shaft axis.

Index

Index

A

Accumulators, 128–129
Acoustical emission monitoring, 51–52
Actuators, 133–134
Air leaks
   in compressors, 125
   in pumps, 122–124
Amplitude of vibration, 52
Arc shields, 143–144

B

Back-up systems, 25
Ballasts, 221
Bearings
   in belt conveyors, 183–184
   in belt drives, 176
   in chain drives, 179
   in compressors, 126
   expansion in, 71, 72

R

S

T